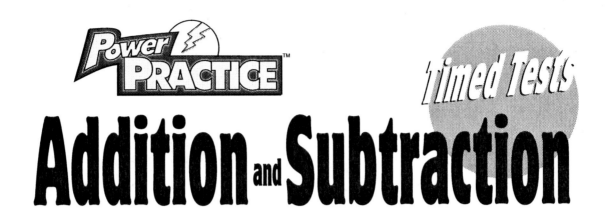

Power PRACTICE™
Addition and Subtraction
Timed Tests

Editor: Carla Hamaguchi

Designer/Production: Moonhee Pak/Rosa Gandara

Cover Designer: Barbara Peterson

Art Director: Tom Cochrane

Project Director: Carolea Williams

© 2004 Creative Teaching Press, Inc., Huntington Beach, CA 92649
Reproduction of activities in any manner for use in the classroom and not for commercial sale is permissible.
Reproduction of these materials for an entire school or for a school system is strictly prohibited.

Table of Contents

Introduction

Each book in the *Power Practice*™ series contains dozens of ready-to-use activity pages to provide students with skill practice. The fun activities can be used to supplement and enhance what you are already teaching in your classroom. Give an activity page to students as independent class work, or send the pages home as homework to reinforce skills taught in class.

The pages in *Timed Tests: Addition and Subtraction* provide students with opportunities to practice and memorize addition and subtraction facts. Challenge students to do their personal best by timing them as they complete each test and keeping track of those times. The tests are intended to be repeated by students until they have reached the desired accuracy and speed.

Have students write a goal time on each test. When they complete a test, have them write their actual time on their test. Encourage students to retake a test to try to better their time and accuracy. Give students record sheets (pages 4 and 63) to keep track of their own scores. An answer key is provided at the back of the book for quick reference.

For extra practice, have students complete the challenge pages. The addition and subtraction challenge pages include problems with missing addends or minuends.

Reward students for earning perfect scores, meeting their personal time goals, or for mastering all the addition and subtraction facts.

Use these motivating timed tests to "recharge" skill review and give students the power to succeed!

Name _____

Addition Record Sheet

Test	Time	Score	Test	Time	Score
Warm-Up +2			Challenge +11		
Practice +2			Warm-Up +12		
Challenge +2			Practice +12		
Warm-Up +3			Challenge +12		
Practice +3			Warm-Up +13		
Challenge +3			Practice +13		
Warm-Up +4			Challenge +13		
Practice +4			Warm-Up +14		
Challenge +4			Practice +14		
Warm-Up +5			Challenge +14		
Practice +5			Warm-Up +15		
Challenge +5			Practice +15		
Warm-Up +6			Challenge +15		
Practice +6			Warm-Up +16		
Challenge +6			Practice +16		
Warm-Up +7			Challenge +16		
Practice +7			Warm-Up +17		
Challenge +7			Practice +17		
Warm-Up +8			Challenge +17		
Practice +8			Warm-Up +18		
Challenge +8			Practice +18		
Warm-Up +9			Challenge +18		
Practice +9			Warm-Up +19		
Challenge +9			Practice +19		
Warm-Up +10			Challenge +19		
Practice +10			Review +1−10		
Challenge +10			Review +11−20		
Warm-Up +11			Review +1−20A		
Practice +11			Review +1−20B		

Timed Tests: Addition and Subtraction © 2004 Creative Teaching Press

Name _____ Date _____

Goal Time _____ Actual Time _____ Score _____

 # Addition Warm-Up +2

1 8
    + 2

2 3
    + 2

3 6
    + 2

4 4
    + 2

5 14
    + 2

6 10
    + 2

7 1
    + 2

8 2
    + 2

9 19
    + 2

10 5
     + 2

11 13
     + 2

12 9
     + 2

13 15
     + 2

14 11
     + 2

15 17
     + 2

16 12
     + 2

17 18
     + 2

18 7
     + 2

19 20
     + 2

20 16
     + 2

Timed Tests: Addition and Subtraction © 2004 Creative Teaching Press

Name _____ Date _____

Goal Time _____ Actual Time _____ Score _____

Addition Practice +2

1 2
    + 5

2 2
   +10

3 2
   + 8

4 2
   +19

5 2
   +15

6 2
    + 3

7 2
    + 1

8 2
   +14

9 2
   +12

10 2
   +18

11 2
    + 6

12 2
   +17

13 2
    + 9

14 2
    + 2

15 2
   +11

16 2
   +16

17 2
    + 4

18 2
   +20

19 2
   +13

20 2
    + 7

Timed Tests: Addition and Subtraction © 2004 Creative Teaching Press

Name _____ Date _____

Goal Time _____   Actual Time _____   Score _____

 # Addition Challenge   +2

1 ☐
+ 2
―――
12

2 ☐
+ 2
―――
5

3 ☐
+ 2
―――
18

4 ☐
+ 2
―――
11

5 ☐
+ 2
―――
16

6 ☐
+ 2
―――
8

7 ☐
+ 2
―――
19

8 ☐
+ 2
―――
3

9 ☐
+ 2
―――
14

10 ☐
+ 2
―――
21

11 ☐
+ 2
―――
9

12 ☐
+ 2
―――
7

13 ☐
+ 2
―――
13

14 ☐
+ 2
―――
10

15 ☐
+ 2
―――
4

16 ☐
+ 2
―――
17

17 ☐
+ 2
―――
20

18 ☐
+ 2
―――
6

19 ☐
+ 2
―――
15

20 ☐
+ 2
―――
22

Timed Tests: Addition and Subtraction © 2004 Creative Teaching Press

Name _____ Date _____

Goal Time _____ Actual Time _____ Score _____

 # Addition Warm-Up   +3

1 3
    + 3

2 13
    + 3

3 6
    + 3

4 1
    + 3

5 14
    + 3

6 10
    + 3

7 7
    + 3

8 4
    + 3

9 19
    + 3

10 9
    + 3

11 5
    + 3

12 8
    + 3

13 16
    + 3

14 11
    + 3

15 17
    + 3

16 12
    + 3

17 20
    + 3

18 18
    + 3

19 2
    + 3

20 15
    + 3

Timed Tests: Addition and Subtraction © 2004 Creative Teaching Press

Name _____ Date _____

Goal Time _____ Actual Time _____ Score _____

 # Addition Practice   +3

1 3
  + 9

2 3
  +10

3 3
  + 3

4 3
  +19

5 3
  +16

6 3
  +14

7 3
  +15

8 3
  +13

9 3
  +12

10 3
  +18

11 3
  + 8

12 3
  + 5

13 3
  +17

14 3
  + 1

15 3
  + 6

16 3
  +11

17 3
  + 2

18 3
  +20

19 3
  + 7

20 3
  + 4

Timed Tests: Addition and Subtraction © 2004 Creative Teaching Press

Addition Challenge   +3

❶
$$\boxed{} \atop {+\ 3 \over 18}$$

❷
$$\boxed{} \atop {+\ 3 \over 10}$$

❸
$$\boxed{} \atop {+\ 3 \over 22}$$

❹
$$\boxed{} \atop {+\ 3 \over 7}$$

❺
$$\boxed{} \atop {+\ 3 \over 14}$$

❻
$$\boxed{} \atop {+\ 3 \over 12}$$

❼
$$\boxed{} \atop {+\ 3 \over 19}$$

❽
$$\boxed{} \atop {+\ 3 \over 4}$$

❾
$$\boxed{} \atop {+\ 3 \over 16}$$

❿
$$\boxed{} \atop {+\ 3 \over 21}$$

⓫
$$\boxed{} \atop {+\ 3 \over 6}$$

⓬
$$\boxed{} \atop {+\ 3 \over 13}$$

⓭
$$\boxed{} \atop {+\ 3 \over 23}$$

⓮
$$\boxed{} \atop {+\ 3 \over 9}$$

⓯
$$\boxed{} \atop {+\ 3 \over 17}$$

⓰
$$\boxed{} \atop {+\ 3 \over 15}$$

⓱
$$\boxed{} \atop {+\ 3 \over 8}$$

⓲
$$\boxed{} \atop {+\ 3 \over 11}$$

⓳
$$\boxed{} \atop {+\ 3 \over 20}$$

⓴
$$\boxed{} \atop {+\ 3 \over 5}$$

Timed Tests: Addition and Subtraction © 2004 Creative Teaching Press

 # Addition Warm-Up   +4

1 15
+ 4

2 3
+ 4

3 12
+ 4

4 19
+ 4

5 6
+ 4

6 2
+ 4

7 17
+ 4

8 10
+ 4

9 8
+ 4

10 14
+ 4

11 13
+ 4

12 1
+ 4

13 5
+ 4

14 16
+ 4

15 9
+ 4

16 20
+ 4

17 7
+ 4

18 11
+ 4

19 18
+ 4

20 4
+ 4

Timed Tests: Addition and Subtraction © 2004 Creative Teaching Press

Name _____ Date _____

Goal Time _____ Actual Time _____ Score _____

Addition Practice   +4

1   4
     +14

2   4
     + 2

3   4
     +19

4   4
     + 8

5   4
     + 9

6   4
     + 4

7   4
     +18

8   4
     +11

9   4
     + 5

10   4
      +15

11   4
      +10

12   4
      + 6

13   4
      +17

14   4
      +12

15   4
      + 1

16   4
      +13

17   4
      +20

18   4
      +16

19   4
      + 3

20   4
      + 7

Timed Tests: Addition and Subtraction © 2004 Creative Teaching Press

Name _____ Date _____

Goal Time _____ Actual Time _____ Score _____

 # Addition Challenge +4

1. □
  + 4
  ――
   7

2 □
  + 4
  ――
  21

3 □
  + 4
  ――
  15

4 □
  + 4
  ――
  11

5 □
  + 4
  ――
  19

6 □
  + 4
  ――
  16

7 □
  + 4
  ――
  13

8 □
  + 4
  ――
  24

9 □
  + 4
  ――
   9

10 □
  + 4
  ――
  22

11 □
  + 4
  ――
  10

12 □
  + 4
  ――
  18

13 □
  + 4
  ――
   5

14 □
  + 4
  ――
  20

15 □
  + 4
  ――
  14

16 □
  + 4
  ――
  17

17 □
  + 4
  ――
   8

18 □
  + 4
  ――
  23

19 □
  + 4
  ――
  12

20 □
  + 4
  ――
   6

Timed Tests: Addition and Subtraction © 2004 Creative Teaching Press

Addition Warm-Up +5

❶ 1
+ 5

❷ 15
+ 5

❸ 11
+ 5

❹ 7
+ 5

❺ 18
+ 5

❻ 17
+ 5

❼ 3
+ 5

❽ 19
+ 5

❾ 13
+ 5

❿ 8
+ 5

⓫ 14
+ 5

⓬ 9
+ 5

⓭ 5
+ 5

⓮ 16
+ 5

⓯ 2
+ 5

⓰ 4
+ 5

⓱ 20
+ 5

⓲ 12
+ 5

⓳ 6
+ 5

⓴ 10
+ 5

Timed Tests: Addition and Subtraction © 2004 Creative Teaching Press

Name _____ Date _____

Goal Time _____ Actual Time _____ Score _____

 # Addition Practice   +5

1   5
     +14

2   5
     + 2

3   5
     +20

4   5
     + 8

5   5
     +16

6   5
     +18

7   5
     + 4

8   5
     +13

9   5
     + 6

10   5
     +10

11   5
     +11

12   5
     + 7

13   5
     +15

14   5
     + 1

15   5
     +12

16   5
     +19

17   5
     + 3

18   5
     + 9

19   5
     +17

20   5
     + 5

Timed Tests: Addition and Subtraction © 2004 Creative Teaching Press

Name _____ Date _____

Goal Time _____ Actual Time _____ Score _____

Addition Challenge   +5

1 □
$$\begin{array}{r} + 5 \\ \hline 10 \end{array}$$

2 □
$$\begin{array}{r} + 5 \\ \hline 21 \end{array}$$

3 □
$$\begin{array}{r} + 5 \\ \hline 8 \end{array}$$

4 □
$$\begin{array}{r} + 5 \\ \hline 24 \end{array}$$

5 □
$$\begin{array}{r} + 5 \\ \hline 15 \end{array}$$

6 □
$$\begin{array}{r} + 5 \\ \hline 18 \end{array}$$

7 □
$$\begin{array}{r} + 5 \\ \hline 23 \end{array}$$

8 □
$$\begin{array}{r} + 5 \\ \hline 12 \end{array}$$

9 □
$$\begin{array}{r} + 5 \\ \hline 25 \end{array}$$

10 □
$$\begin{array}{r} + 5 \\ \hline 6 \end{array}$$

11 □
$$\begin{array}{r} + 5 \\ \hline 16 \end{array}$$

12 □
$$\begin{array}{r} + 5 \\ \hline 7 \end{array}$$

13 □
$$\begin{array}{r} + 5 \\ \hline 19 \end{array}$$

14 □
$$\begin{array}{r} + 5 \\ \hline 22 \end{array}$$

15 □
$$\begin{array}{r} + 5 \\ \hline 13 \end{array}$$

16 □
$$\begin{array}{r} + 5 \\ \hline 9 \end{array}$$

17 □
$$\begin{array}{r} + 5 \\ \hline 14 \end{array}$$

18 □
$$\begin{array}{r} + 5 \\ \hline 20 \end{array}$$

19 □
$$\begin{array}{r} + 5 \\ \hline 11 \end{array}$$

20 □
$$\begin{array}{r} + 5 \\ \hline 17 \end{array}$$

Timed Tests: Addition and Subtraction © 2004 Creative Teaching Press

Name _____ Date _____

Goal Time _____ Actual Time _____ Score _____

 # Addition Warm-Up   +6

1   9
    + 6

2   18
    + 6

3   2
    + 6

4   14
    + 6

5   8
    + 6

6   20
    + 6

7   1
    + 6

8   16
    + 6

9   5
    + 6

10 12
    + 6

11   4
    + 6

12 13
    + 6

13   7
    + 6

14 19
    + 6

15 10
    + 6

16 17
    + 6

17   6
    + 6

18 11
    + 6

19 15
    + 6

20   3
    + 6

Timed Tests: Addition and Subtraction © 2004 Creative Teaching Press

Name _____ Date _____

Goal Time _____ Actual Time _____ Score _____

Addition Practice   +6

1   6
     +14

2   6
     +18

3   6
     + 3

4   6
     +10

5   6
     + 6

6   6
     + 1

7   6
     +12

8   6
     + 5

9   6
     + 8

10   6
      +17

11   6
      +16

12   6
      +19

13   6
      + 2

14   6
      + 7

15   6
      +11

16   6
      + 4

17   6
      +15

18   6
      + 20

19   6
      + 9

20   6
      +13

18

Timed Tests: Addition and Subtraction © 2004 Creative Teaching Press

Name _____ Date _____

Goal Time _____ Actual Time _____ Score _____

 # Addition Challenge   +6

1 ☐
+ 6
 7

2 ☐
+ 6
 18

3 ☐
+ 6
 12

4 ☐
+ 6
 26

5 ☐
+ 6
 15

6 ☐
+ 6
 13

7 ☐
+ 6
 21

8 ☐
+ 6
 17

9 ☐
+ 6
 10

10 ☐
+ 6
 23

11 ☐
+ 6
 24

12 ☐
+ 6
 9

13 ☐
+ 6
 22

14 ☐
+ 6
 16

15 ☐
+ 6
 8

16 ☐
+ 6
 11

17 ☐
+ 6
 19

18 ☐
+ 6
 25

19 ☐
+ 6
 14

20 ☐
+ 6
 20

Timed Tests: Addition and Subtraction © 2004 Creative Teaching Press

Name _____ Date _____

Goal Time _____ Actual Time _____ Score _____

 # Addition Warm-Up   +7

1 $\begin{array}{r} 2 \\ +\ 7 \\ \hline \end{array}$ **2** $\begin{array}{r} 10 \\ +\ 7 \\ \hline \end{array}$ **3** $\begin{array}{r} 19 \\ +\ 7 \\ \hline \end{array}$ **4** $\begin{array}{r} 5 \\ +\ 7 \\ \hline \end{array}$ **5** $\begin{array}{r} 13 \\ +\ 7 \\ \hline \end{array}$

6 $\begin{array}{r} 14 \\ +\ 7 \\ \hline \end{array}$ **7** $\begin{array}{r} 6 \\ +\ 7 \\ \hline \end{array}$ **8** $\begin{array}{r} 1 \\ +\ 7 \\ \hline \end{array}$ **9** $\begin{array}{r} 17 \\ +\ 7 \\ \hline \end{array}$ **10** $\begin{array}{r} 8 \\ +\ 7 \\ \hline \end{array}$

11 $\begin{array}{r} 7 \\ +\ 7 \\ \hline \end{array}$ **12** $\begin{array}{r} 18 \\ +\ 7 \\ \hline \end{array}$ **13** $\begin{array}{r} 9 \\ +\ 7 \\ \hline \end{array}$ **14** $\begin{array}{r} 11 \\ +\ 7 \\ \hline \end{array}$ **15** $\begin{array}{r} 3 \\ +\ 7 \\ \hline \end{array}$

16 $\begin{array}{r} 16 \\ +\ 7 \\ \hline \end{array}$ **17** $\begin{array}{r} 4 \\ +\ 7 \\ \hline \end{array}$ **18** $\begin{array}{r} 12 \\ +\ 7 \\ \hline \end{array}$ **19** $\begin{array}{r} 20 \\ +\ 7 \\ \hline \end{array}$ **20** $\begin{array}{r} 15 \\ +\ 7 \\ \hline \end{array}$

Timed Tests: Addition and Subtraction © 2004 Creative Teaching Press

Name _____ Date _____

Goal Time _____ Actual Time _____ Score _____

Addition Practice +7

1 $\begin{array}{r} 7 \\ +\ 9 \\ \hline \end{array}$ **2** $\begin{array}{r} 7 \\ +13 \\ \hline \end{array}$ **3** $\begin{array}{r} 7 \\ +\ 5 \\ \hline \end{array}$ **4** $\begin{array}{r} 7 \\ +16 \\ \hline \end{array}$ **5** $\begin{array}{r} 7 \\ +11 \\ \hline \end{array}$

6 $\begin{array}{r} 7 \\ +\ 6 \\ \hline \end{array}$ **7** $\begin{array}{r} 7 \\ +17 \\ \hline \end{array}$ **8** $\begin{array}{r} 7 \\ +\ 8 \\ \hline \end{array}$ **9** $\begin{array}{r} 7 \\ +\ 2 \\ \hline \end{array}$ **10** $\begin{array}{r} 7 \\ +20 \\ \hline \end{array}$

11 $\begin{array}{r} 7 \\ +18 \\ \hline \end{array}$ **12** $\begin{array}{r} 7 \\ +\ 3 \\ \hline \end{array}$ **13** $\begin{array}{r} 7 \\ +10 \\ \hline \end{array}$ **14** $\begin{array}{r} 7 \\ +15 \\ \hline \end{array}$ **15** $\begin{array}{r} 7 \\ +\ 4 \\ \hline \end{array}$

16 $\begin{array}{r} 7 \\ +\ 1 \\ \hline \end{array}$ **17** $\begin{array}{r} 7 \\ +14 \\ \hline \end{array}$ **18** $\begin{array}{r} 7 \\ +\ 7 \\ \hline \end{array}$ **19** $\begin{array}{r} 7 \\ +19 \\ \hline \end{array}$ **20** $\begin{array}{r} 7 \\ +12 \\ \hline \end{array}$

Timed Tests: Addition and Subtraction © 2004 Creative Teaching Press

Addition Challenge   +7

1 ▢
+ 7
22

2 ▢
+ 7
13

3 ▢
+ 7
25

4 ▢
+ 7
18

5 ▢
+ 7
10

6 ▢
+ 7
16

7 ▢
+ 7
8

8 ▢
+ 7
20

9 ▢
+ 7
12

10 ▢
+ 7
26

11 ▢
+ 7
24

12 ▢
+ 7
17

13 ▢
+ 7
14

14 ▢
+ 7
9

15 ▢
+ 7
21

16 ▢
+ 7
11

17 ▢
+ 7
27

18 ▢
+ 7
19

19 ▢
+ 7
23

20 ▢
+ 7
15

Timed Tests: Addition and Subtraction © 2004 Creative Teaching Press

Goal Time _____ Actual Time _____ Score _____

Addition Warm-Up +8

1 7
  + 8

2 12
  + 8

3 9
  + 8

4 16
  + 8

5 1
  + 8

6 19
  + 8

7 2
  + 8

8 14
  + 8

9 6
  + 8

10 17
  + 8

11 4
  + 8

12 10
  + 8

13 18
  + 8

14 3
  + 8

15 15
  + 8

16 13
  + 8

17 5
  + 8

18 20
  + 8

19 8
  + 8

20 11
  + 8

Timed Tests: Addition and Subtraction © 2004 Creative Teaching Press

Addition Practice   +8

❶ $\begin{array}{r} 8 \\ +\ 2 \\ \hline \end{array}$ ❷ $\begin{array}{r} 8 \\ +16 \\ \hline \end{array}$ ❸ $\begin{array}{r} 8 \\ +\ 8 \\ \hline \end{array}$ ❹ $\begin{array}{r} 8 \\ +\ 4 \\ \hline \end{array}$ ❺ $\begin{array}{r} 8 \\ +19 \\ \hline \end{array}$

❻ $\begin{array}{r} 8 \\ +\ 7 \\ \hline \end{array}$ ❼ $\begin{array}{r} 8 \\ +11 \\ \hline \end{array}$ ❽ $\begin{array}{r} 8 \\ +\ 5 \\ \hline \end{array}$ ❾ $\begin{array}{r} 8 \\ +14 \\ \hline \end{array}$ ❿ $\begin{array}{r} 8 \\ +18 \\ \hline \end{array}$

⓫ $\begin{array}{r} 8 \\ +\ 1 \\ \hline \end{array}$ ⓬ $\begin{array}{r} 8 \\ +\ 9 \\ \hline \end{array}$ ⓭ $\begin{array}{r} 8 \\ +20 \\ \hline \end{array}$ ⓮ $\begin{array}{r} 8 \\ +12 \\ \hline \end{array}$ ⓯ $\begin{array}{r} 8 \\ +\ 6 \\ \hline \end{array}$

⓰ $\begin{array}{r} 8 \\ +13 \\ \hline \end{array}$ ⓱ $\begin{array}{r} 8 \\ +17 \\ \hline \end{array}$ ⓲ $\begin{array}{r} 8 \\ +\ 3 \\ \hline \end{array}$ ⓳ $\begin{array}{r} 8 \\ +15 \\ \hline \end{array}$ ⓴ $\begin{array}{r} 8 \\ +10 \\ \hline \end{array}$

Timed Tests: Addition and Subtraction © 2004 Creative Teaching Press

Name _____ Date _____

Goal Time _____ Actual Time _____ Score _____

Addition Challenge +8

1 ☐
  + 8
  16

2 ☐
  + 8
  23

3 ☐
  + 8
  10

4 ☐
  + 8
  26

5 ☐
  + 8
  15

6 ☐
  + 8
  18

7 ☐
  + 8
  14

8 ☐
  + 8
  27

9 ☐
  + 8
  12

10 ☐
  + 8
  19

11 ☐
  + 8
  21

12 ☐
  + 8
  9

13 ☐
  + 8
  24

14 ☐
  + 8
  17

15 ☐
  + 8
  28

16 ☐
  + 8
  25

17 ☐
  + 8
  13

18 ☐
  + 8
  20

19 ☐
  + 8
  11

20 ☐
  + 8
  22

Timed Tests: Addition and Subtraction © 2004 Creative Teaching Press

Name _____ Date _____

Goal Time _____ Actual Time _____ Score _____

 # Addition Warm-Up +9

1 10
    + 9

2 13
    + 9

3 4
    + 9

4 18
    + 9

5 6
    + 9

6 7
    + 9

7 16
    + 9

8 1
    + 9

9 9
    + 9

10 14
     + 9

11 2
     + 9

12 19
     + 9

13 11
     + 9

14 5
     + 9

15 17
     + 9

16 8
     + 9

17 15
     + 9

18 3
     + 9

19 20
     + 9

20 12
     + 9

Timed Tests: Addition and Subtraction © 2004 Creative Teaching Press

Name _____ Date _____

Goal Time _____ Actual Time _____ Score _____

Addition Practice   +9

1 9 +20

2 9 + 8

3 9 +16

4 9 + 3

5 9 +19

6 9 +12

7 9 +14

8 9 + 5

9 9 +10

10 9 +17

11 9 +18

12 9 + 2

13 9 + 9

14 9 + 6

15 9 +13

16 9 + 4

17 9 +11

18 9 + 7

19 9 +15

20 9 + 1

Timed Tests: Addition and Subtraction © 2004 Creative Teaching Press

Addition Challenge   +9

1 ☐
+ 9
14

2 ☐
+ 9
22

3 ☐
+ 9
20

4 ☐
+ 9
11

5 ☐
+ 9
27

6 ☐
+ 9
16

7 ☐
+ 9
24

8 ☐
+ 9
12

9 ☐
+ 9
29

10 ☐
+ 9
18

11 ☐
+ 9
25

12 ☐
+ 9
10

13 ☐
+ 9
21

14 ☐
+ 9
15

15 ☐
+ 9
23

16 ☐
+ 9
19

17 ☐
+ 9
17

18 ☐
+ 9
26

19 ☐
+ 9
13

20 ☐
+ 9
28

Timed Tests: Addition and Subtraction © 2004 Creative Teaching Press

Name _____   Date _____

Goal Time _____   Actual Time _____   Score _____

 # Addition Warm-Up   +10

1 3
  +10

2 17
  +10

3 8
  +10

4 20
  +10

5 13
  +10

6 6
  +10

7 4
  +10

8 18
  +10

9 12
  +10

10 7
  +10

11 14
  +10

12 10
  +10

13 2
  +10

14 5
  +10

15 19
  +10

16 1
  +10

17 11
  +10

18 9
  +10

19 16
  +10

20 15
  +10

Timed Tests: Addition and Subtraction © 2004 Creative Teaching Press

Goal Time _____   Actual Time _____   Score _____

Addition Practice   +10

① 10
  + 9

② 10
  + 7

③ 10
  +12

④ 10
  +18

⑤ 10
  + 3

⑥ 10
  + 4

⑦ 10
  +19

⑧ 10
  + 1

⑨ 10
  +15

⑩ 10
  +13

⑪ 10
  +14

⑫ 10
  + 2

⑬ 10
  +11

⑭ 10
  + 8

⑮ 10
  +16

⑯ 10
  + 5

⑰ 10
  +10

⑱ 10
  + 6

⑲ 10
  +17

⑳ 10
  +20

Timed Tests: Addition and Subtraction © 2004 Creative Teaching Press

Name _____ Date _____

Goal Time _____ Actual Time _____ Score _____

 # Addition Challenge +10

1
☐
+10
—
21

2
☐
+10
—
17

3
☐
+10
—
28

4
☐
+10
—
15

5
☐
+10
—
23

6
☐
+10
—
14

7
☐
+10
—
29

8
☐
+10
—
12

9
☐
+10
—
25

10
☐
+10
—
11

11
☐
+10
—
19

12
☐
+10
—
24

13
☐
+10
—
18

14
☐
+10
—
30

15
☐
+10
—
26

16
☐
+10
—
22

17
☐
+10
—
16

18
☐
+10
—
27

19
☐
+10
—
13

20
☐
+10
—
20

Timed Tests: Addition and Subtraction © 2004 Creative Teaching Press

Name _____   Date _____

Goal Time _____   Actual Time _____   Score _____

Addition Warm-Up   +11

❶   6
     +11

❷   20
    +11

❸   4
     +11

❹   17
    +11

❺   14
    +11

❻   19
    +11

❼   8
     +11

❽   1
     +11

❾   12
    +11

❿   10
    +11

⓫   2
     +11

⓬   15
    +11

⓭   9
     +11

⓮   5
     +11

⓯   13
    +11

⓰   11
    +11

⓱   3
     +11

⓲   18
    +11

⓳   7
     +11

⓴   16
    +11

Timed Tests: Addition and Subtraction © 2004 Creative Teaching Press

Name _____   Date _____

Goal Time _____   Actual Time _____   Score _____

 # Addition Practice   +11

1 11
    + 5

2 11
    +10

3 11
    + 8

4 11
    +19

5 11
    +13

6 11
    +16

7 11
    + 3

8 11
    +14

9 11
    + 6

10 11
     +11

11 11
     + 4

12 11
     + 9

13 11
     +12

14 11
     + 1

15 11
     +20

16 11
     +18

17 11
     + 2

18 11
     +15

19 11
     + 7

20 11
     +17

Timed Tests: Addition and Subtraction © 2004 Creative Teaching Press

Name _____ Date _____

Goal Time _____ Actual Time _____ Score _____

 # Addition Challenge   +11

❶ ☐
+11
22

❷ ☐
+11
30

❸ ☐
+11
12

❹ ☐
+11
25

❺ ☐
+11
19

❻ ☐
+11
16

❼ ☐
+11
26

❽ ☐
+11
18

❾ ☐
+11
31

❿ ☐
+11
15

⓫ ☐
+11
29

⓬ ☐
+11
14

⓭ ☐
+11
21

⓮ ☐
+11
27

⓯ ☐
+11
23

⓰ ☐
+11
20

⓱ ☐
+11
17

⓲ ☐
+11
24

⓳ ☐
+11
13

⓴ ☐
+11
28

Timed Tests: Addition and Subtraction © 2004 Creative Teaching Press

Name _____ Date _____

Goal Time _____ Actual Time _____ Score _____

 **Addition Warm-Up   +12**

❶
$\begin{array}{r} 3 \\ +12 \\ \hline \end{array}$
❷
$\begin{array}{r} 11 \\ +12 \\ \hline \end{array}$
❸
$\begin{array}{r} 17 \\ +12 \\ \hline \end{array}$
❹
$\begin{array}{r} 6 \\ +12 \\ \hline \end{array}$
❺
$\begin{array}{r} 14 \\ +12 \\ \hline \end{array}$

❻
$\begin{array}{r} 19 \\ +12 \\ \hline \end{array}$
❼
$\begin{array}{r} 7 \\ +12 \\ \hline \end{array}$
❽
$\begin{array}{r} 2 \\ +12 \\ \hline \end{array}$
❾
$\begin{array}{r} 15 \\ +12 \\ \hline \end{array}$
❿
$\begin{array}{r} 20 \\ +12 \\ \hline \end{array}$

⓫
$\begin{array}{r} 5 \\ +12 \\ \hline \end{array}$
⓬
$\begin{array}{r} 12 \\ +12 \\ \hline \end{array}$
⓭
$\begin{array}{r} 9 \\ +12 \\ \hline \end{array}$
⓮
$\begin{array}{r} 18 \\ +12 \\ \hline \end{array}$
⓯
$\begin{array}{r} 10 \\ +12 \\ \hline \end{array}$

⓰
$\begin{array}{r} 1 \\ +12 \\ \hline \end{array}$
⓱
$\begin{array}{r} 8 \\ +12 \\ \hline \end{array}$
⓲
$\begin{array}{r} 16 \\ +12 \\ \hline \end{array}$
⓳
$\begin{array}{r} 13 \\ +12 \\ \hline \end{array}$
⓴
$\begin{array}{r} 4 \\ +12 \\ \hline \end{array}$

Timed Tests: Addition and Subtraction © 2004 Creative Teaching Press

Addition Practice   +12

1 12
  +18

2 12
  + 8

3 12
  +13

4 12
  + 4

5 12
  +20

6 12
  + 5

7 12
  +12

8 12
  +10

9 12
  +16

10 12
  + 1

11 12
  +14

12 12
  + 2

13 12
  +17

14 12
  + 6

15 12
  +19

16 12
  + 7

17 12
  +11

18 12
  + 3

19 12
  +15

20 12
  + 9

Timed Tests: Addition and Subtraction © 2004 Creative Teaching Press

Name _____ Date _____

Goal Time _____ Actual Time _____ Score _____

 # Addition Challenge +12

❶ ☐
+12
24

❷ ☐
+12
31

❸ ☐
+12
13

❹ ☐
+12
27

❺ ☐
+12
17

❻ ☐
+12
14

❼ ☐
+12
23

❽ ☐
+12
18

❾ ☐
+12
29

❿ ☐
+12
20

⓫ ☐
+12
22

⓬ ☐
+12
28

⓭ ☐
+12
32

⓮ ☐
+12
25

⓯ ☐
+12
15

⓰ ☐
+12
26

⓱ ☐
+12
21

⓲ ☐
+12
16

⓳ ☐
+12
30

⓴ ☐
+12
19

Timed Tests: Addition and Subtraction © 2004 Creative Teaching Press

Name _____ Date _____

Goal Time _____ Actual Time _____ Score _____

Addition Warm-Up   +13

1   6   **2**   19   **3**   5   **4**   11   **5**   16

    +13   +13   +13   +13   +13

6   13   **7**   4   **8**   20   **9**   9   **10**   2

    +13   +13   +13   +13   +13

11   8   **12**   15   **13**   1   **14**   14   **15**   7

    +13   +13   +13   +13   +13

16   3   **17**   10   **18**   17   **19**   12   **20**   18

    +13   +13   +13   +13   +13

Timed Tests: Addition and Subtraction © 2004 Creative Teaching Press

Addition Practice +13

1 13
    +11

2 13
    + 3

3 13
    +16

4 13
    +20

5 13
    + 6

6 13
    + 5

7 13
    + 8

8 13
    +17

9 13
    + 1

10 13
     +12

11 13
     +10

12 13
     + 4

13 13
     +19

14 13
     + 7

15 13
     + 9

16 13
     +14

17 13
     + 2

18 13
     +18

19 13
     +15

20 13
     +13

Timed Tests: Addition and Subtraction © 2004 Creative Teaching Press

Name _____ Date _____

Goal Time _____ Actual Time _____ Score _____

 **Addition Challenge   +13**

1 ☐
+13
14

2 ☐
+13
22

3 ☐
+13
30

4 ☐
+13
18

5 ☐
+13
26

6 ☐
+13
28

7 ☐
+13
17

8 ☐
+13
32

9 ☐
+13
21

10 ☐
+13
24

11 ☐
+13
25

12 ☐
+13
20

13 ☐
+13
15

14 ☐
+13
27

15 ☐
+13
31

16 ☐
+13
16

17 ☐
+13
29

18 ☐
+13
23

19 ☐
+13
33

20 ☐
+13
19

Timed Tests: Addition and Subtraction © 2004 Creative Teaching Press

Name _____ Date _____

Goal Time _____ Actual Time _____ Score _____

 # Addition Warm-Up   +14

1 12 +14

2 6 +14

3 19 +14

4 2 +14

5 15 +14

6 4 +14

7 20 +14

8 7 +14

9 18 +14

10 10 +14

11 17 +14

12 8 +14

13 11 +14

14 1 +14

15 13 +14

16 9 +14

17 3 +14

18 14 +14

19 5 +14

20 16 +14

Timed Tests: Addition and Subtraction © 2004 Creative Teaching Press

Addition Practice   +14

1 14
    + 9

2 14
    +11

3 14
    + 8

4 14
    + 3

5 14
    +18

6 14
    + 1

7 14
    +16

8 14
    + 5

9 14
    +12

10 14
     +19

11 14
     + 7

12 14
     +13

13 14
     + 2

14 14
     + 6

15 14
     +10

16 14
     + 4

17 14
     +14

18 14
     +15

19 14
     +20

20 14
     +17

Timed Tests: Addition and Subtraction © 2004 Creative Teaching Press

Name _____ Date _____

Goal Time _____ Actual Time _____ Score _____

 # Addition Challenge   +14

❶ ☐
+14
30

❷ ☐
+14
17

❸ ☐
+14
28

❹ ☐
+14
20

❺ ☐
+14
25

❻ ☐
+14
19

❼ ☐
+14
23

❽ ☐
+14
31

❾ ☐
+14
15

❿ ☐
+14
33

⓫ ☐
+14
27

⓬ ☐
+14
18

⓭ ☐
+14
24

⓮ ☐
+14
32

⓯ ☐
+14
21

⓰ ☐
+14
26

⓱ ☐
+14
34

⓲ ☐
+14
16

⓳ ☐
+14
22

⓴ ☐
+14
29

Timed Tests: Addition and Subtraction © 2004 Creative Teaching Press

Addition Warm-Up   +15

❶ 5
 +15

❷ 16
 +15

❸ 3
 +15

❹ 8
 +15

❺ 10
 +15

❻ 13
 +15

❼ 6
 +15

❽ 18
 +15

❾ 20
 +15

❿ 14
 +15

⓫ 11
 +15

⓬ 2
 +15

⓭ 15
 +15

⓮ 9
 +15

⓯ 17
 +15

⓰ 1
 +15

⓱ 19
 +15

⓲ 7
 +15

⓳ 4
 +15

⓴ 12
 +15

Timed Tests: Addition and Subtraction © 2004 Creative Teaching Press

Addition Practice   +15

1 15
    + 9

2 15
    + 3

3 15
    +14

4 15
    + 1

5 15
    + 8

6 15
    +12

7 15
    + 7

8 15
    +16

9 15
    + 5

10 15
    +19

11 15
    + 6

12 15
    +10

13 15
    + 2

14 15
    +13

15 15
    +17

16 15
    +15

17 15
    + 4

18 15
    +18

19 15
    +11

20 15
    +20

Timed Tests: Addition and Subtraction © 2004 Creative Teaching Press

Name _____ Date _____

Goal Time _____ Actual Time _____ Score _____

 # Addition Challenge   +15

1 ☐
  +15
  26

2 ☐
  +15
  22

3 ☐
  +15
  31

4 ☐
  +15
  18

5 ☐
  +15
  29

6 ☐
  +15
  17

7 ☐
  +15
  32

8 ☐
  +15
  20

9 ☐
  +15
  35

10 ☐
  +15
  24

11 ☐
  +15
  21

12 ☐
  +15
  30

13 ☐
  +15
  25

14 ☐
  +15
  16

15 ☐
  +15
  27

16 ☐
  +15
  28

17 ☐
  +15
  19

18 ☐
  +15
  34

19 ☐
  +15
  23

20 ☐
  +15
  33

Timed Tests: Addition and Subtraction © 2004 Creative Teaching Press

Name _____ Date _____

Goal Time _____ Actual Time _____ Score _____

 ## Addition Warm-Up   +16

1 1
    +16

2 14
    +16

3 7
    +16

4 18
    +16

5 5
    +16

6 12
    +16

7 9
    +16

8 4
    +16

9 15
    +16

10 20
     +16

11 6
     +16

12 17
     +16

13 8
     +16

14 10
     +16

15 2
     +16

16 3
     +16

17 16
     +16

18 11
     +16

19 13
     +16

20 19
     +16

Timed Tests: Addition and Subtraction © 2004 Creative Teaching Press

Addition Practice   +16

1 16
   + 9

2 16
   + 3

3 16
   +16

4 16
   +18

5 16
   + 5

6 16
   + 8

7 16
   + 6

8 16
   +12

9 16
   + 4

10 16
    +10

11 16
    +14

12 16
    + 1

13 16
    +17

14 16
    + 7

15 16
    +13

16 16
    +11

17 16
    +20

18 16
    +15

19 16
    + 2

20 16
    +19

Timed Tests: Addition and Subtraction © 2004 Creative Teaching Press

Name _____ Date _____

Goal Time _____ Actual Time _____ Score _____

 # Addition Challenge   +16

1 ☐
+16
30

2 ☐
+16
32

3 ☐
+16
19

4 ☐
+16
35

5 ☐
+16
26

6 ☐
+16
20

7 ☐
+16
27

8 ☐
+16
36

9 ☐
+16
23

10 ☐
+16
31

11 ☐
+16
24

12 ☐
+16
18

13 ☐
+16
34

14 ☐
+16
21

15 ☐
+16
28

16 ☐
+16
29

17 ☐
+16
22

18 ☐
+16
33

19 ☐
+16
25

20 ☐
+16
17

Timed Tests: Addition and Subtraction © 2004 Creative Teaching Press

Addition Warm-Up   +17

① 5 +17

② 15 +17

③ 10 +17

④ 8 +17

⑤ 13 +17

⑥ 16 +17

⑦ 4 +17

⑧ 19 +17

⑨ 2 +17

⑩ 20 +17

⑪ 9 +17

⑫ 1 +17

⑬ 18 +17

⑭ 6 +17

⑮ 11 +17

⑯ 3 +17

⑰ 12 +17

⑱ 7 +17

⑲ 14 +17

⑳ 17 +17

Timed Tests: Addition and Subtraction © 2004 Creative Teaching Press

Name _____ Date _____

Goal Time _____ Actual Time _____ Score _____

Addition Practice +17

❶ 17
+ 2

❷ 17
+13

❸ 17
+20

❹ 17
+ 4

❺ 17
+17

❻ 17
+ 6

❼ 17
+ 9

❽ 17
+ 1

❾ 17
+19

❿ 17
+14

⓫ 17
+12

⓬ 17
+ 7

⓭ 17
+ 3

⓮ 17
+15

⓯ 17
+10

⓰ 17
+16

⓱ 17
+ 5

⓲ 17
+11

⓳ 17
+ 8

⓴ 17
+18

Timed Tests: Addition and Subtraction © 2004 Creative Teaching Press

Addition Challenge +17

1 ☐
+17
20

2 ☐
+17
26

3 ☐
+17
34

4 ☐
+17
23

5 ☐
+17
31

6 ☐
+17
28

7 ☐
+17
22

8 ☐
+17
35

9 ☐
+17
30

10 ☐
+17
18

11 ☐
+17
32

12 ☐
+17
19

13 ☐
+17
37

14 ☐
+17
25

15 ☐
+17
36

16 ☐
+17
24

17 ☐
+17
27

18 ☐
+17
33

19 ☐
+17
21

20 ☐
+17
29

Timed Tests: Addition and Subtraction © 2004 Creative Teaching Press

Name _____ Date _____

Goal Time _____ Actual Time _____ Score _____

 # Addition Warm-Up   +18

1 9
   +18

2 5
   +18

3 12
   +18

4 1
   +18

5 16
   +18

6 14
   +18

7 2
   +18

8 20
   +18

9 7
   +18

10 18
    +18

11 4
    +18

12 17
    +18

13 8
    +18

14 13
    +18

15 10
    +18

16 19
    +18

17 6
    +18

18 11
    +18

19 3
    +18

20 15
    +18

Timed Tests: Addition and Subtraction © 2004 Creative Teaching Press

Name _____ Date _____

Goal Time _____ Actual Time _____ Score _____

 # Addition Practice   +18

1 18
+ 6

2 18
+15

3 18
+ 2

4 18
+13

5 18
+19

6 18
+ 7

7 18
+14

8 18
+17

9 18
+ 3

10 18
+12

11 18
+20

12 18
+11

13 18
+16

14 18
+ 5

15 18
+18

16 18
+ 9

17 18
+ 1

18 18
+10

19 18
+ 4

20 18
+ 8

Timed Tests: Addition and Subtraction © 2004 Creative Teaching Press

Goal Time _____ Actual Time _____ Score _____

 **Addition Challenge   +18**

1 ☐
$\begin{array}{r} +18 \\ \hline 32 \end{array}$

2 ☐
$\begin{array}{r} +18 \\ \hline 21 \end{array}$

3 ☐
$\begin{array}{r} +18 \\ \hline 36 \end{array}$

4 ☐
$\begin{array}{r} +18 \\ \hline 27 \end{array}$

5 ☐
$\begin{array}{r} +18 \\ \hline 19 \end{array}$

6 ☐
$\begin{array}{r} +18 \\ \hline 23 \end{array}$

7 ☐
$\begin{array}{r} +18 \\ \hline 30 \end{array}$

8 ☐
$\begin{array}{r} +18 \\ \hline 25 \end{array}$

9 ☐
$\begin{array}{r} +18 \\ \hline 38 \end{array}$

10 ☐
$\begin{array}{r} +18 \\ \hline 33 \end{array}$

11 ☐
$\begin{array}{r} +18 \\ \hline 35 \end{array}$

12 ☐
$\begin{array}{r} +18 \\ \hline 20 \end{array}$

13 ☐
$\begin{array}{r} +18 \\ \hline 37 \end{array}$

14 ☐
$\begin{array}{r} +18 \\ \hline 24 \end{array}$

15 ☐
$\begin{array}{r} +18 \\ \hline 29 \end{array}$

16 ☐
$\begin{array}{r} +18 \\ \hline 28 \end{array}$

17 ☐
$\begin{array}{r} +18 \\ \hline 34 \end{array}$

18 ☐
$\begin{array}{r} +18 \\ \hline 22 \end{array}$

19 ☐
$\begin{array}{r} +18 \\ \hline 31 \end{array}$

20 ☐
$\begin{array}{r} +18 \\ \hline 26 \end{array}$

Timed Tests: Addition and Subtraction © 2004 Creative Teaching Press

Name _____ Date _____

Addition Warm-Up   +19

1 5
+19

2 14
+19

3 7
+19

4 10
+19

5 2
+19

6 1
+19

7 20
+19

8 13
+19

9 9
+19

10 16
+19

11 12
+19

12 3
+19

13 18
+19

14 6
+19

15 11
+19

16 8
+19

17 15
+19

18 19
+19

19 4
+19

20 17
+19

Timed Tests: Addition and Subtraction © 2004 Creative Teaching Press

Name _____ Date _____

Goal Time _____ Actual Time _____ Score _____

 # Addition Practice   +19

1 19
  + 3

2 19
  +11

3 19
  + 8

4 19
  +14

5 19
  +18

6 19
  + 6

7 19
  +15

8 19
  + 4

9 19
  +19

10 19
  +10

11 19
  + 1

12 19
  +12

13 19
  + 20

14 19
  + 5

15 19
  +16

16 19
  +13

17 19
  + 9

18 19
  +17

19 19
  + 2

20 19
  + 7

Timed Tests: Addition and Subtraction © 2004 Creative Teaching Press

Name _____ Date _____

Goal Time _____ Actual Time _____ Score _____

 # Addition Challenge   +19

1 ☐
+19
31

2 ☐
+19
20

3 ☐
+19
35

4 ☐
+19
24

5 ☐
+19
38

6 ☐
+19
22

7 ☐
+19
30

8 ☐
+19
28

9 ☐
+19
33

10 ☐
+19
26

11 ☐
+19
27

12 ☐
+19
34

13 ☐
+19
21

14 ☐
+19
36

15 ☐
+19
32

16 ☐
+19
37

17 ☐
+19
25

18 ☐
+19
39

19 ☐
+19
29

20 ☐
+19
23

Timed Tests: Addition and Subtraction © 2004 Creative Teaching Press

Name _____ Date _____

Goal Time _____ Actual Time _____ Score _____

 # Addition Review   +1–10

1 2
   + 8

2 4
   + 7

3 3
   + 3

4 10
   + 4

5 6
   + 6

6 9
   + 3

7 10
   +10

8 3
   + 4

9 2
   + 6

10 3
   + 5

11 5
   + 4

12 4
   + 8

13 7
   + 7

14 5
   + 1

15 10
   + 8

16 2
   + 9

17 6
   + 7

18 8
   + 9

19 9
   + 6

20 4
   + 6

Timed Tests: Addition and Subtraction © 2004 Creative Teaching Press

Addition Review +11–20

1
```
  19
+12
```
2
```
  14
+16
```
3
```
  11
+14
```
4
```
  17
+16
```
5
```
  15
+13
```

6
```
  13
+18
```
7
```
  16
+16
```
8
```
  19
+17
```
9
```
  15
+16
```
10
```
  20
+14
```

11
```
  11
+13
```
12
```
  17
+14
```
13
```
  12
+16
```
14
```
  20
+15
```
15
```
  12
+19
```

16
```
  19
+18
```
17
```
  14
+15
```
18
```
  11
+18
```
19
```
  13
+16
```
20
```
  19
+16
```

Timed Tests: Addition and Subtraction © 2004 Creative Teaching Press

Name _____ Date _____

Goal Time _____ Actual Time _____ Score _____

 # Addition Review +1–20A

1 6
+ 9

2 10
+ 3

3 20
+ 6

4 8
+ 2

5 13
+ 5

6 16
+ 7

7 15
+ 4

8 19
+11

9 16
+ 4

10 9
+ 5

11 11
+ 5

12 7
+ 7

13 4
+12

14 17
+ 1

15 3
+13

16 8
+ 9

17 2
+20

18 17
+ 8

19 16
+ 1

20 6
+19

Timed Tests: Addition and Subtraction © 2004 Creative Teaching Press

Name _____ Date _____

Goal Time _____ Actual Time _____ Score _____

 # Addition Review   +1–20B

1
2
+ 9

2
8
+ 7

3
18
+ 4

4
6
+11

5
3
+ 5

6
15
+ 8

7
12
+10

8
13
+ 1

9
7
+ 20

10
17
+ 6

11
14
+ 7

12
16
+ 2

13
19
+19

14
4
+ 9

15
15
+ 3

16
20
+19

17
10
+10

18
14
+14

19
18
+ 9

20
6
+ 6

Timed Tests: Addition and Subtraction © 2004 Creative Teaching Press

Name _____

Subtraction Record Sheet

Test	Time	Score	Test	Time	Score
Warm-Up –2			Warm-Up –12		
Practice –2			Practice –12		
Challenge –2			Challenge –12		
Warm-Up –3			Warm-Up –13		
Practice –3			Practice –13		
Challenge –3			Challenge –13		
Warm-Up –4			Warm-Up –14		
Practice –4			Practice –14		
Challenge –4			Challenge –14		
Warm-Up –5			Warm-Up –15		
Practice –5			Practice –15		
Challenge –5			Challenge –15		
Warm-Up –6			Warm-Up –16		
Practice –6			Practice –16		
Challenge –6			Challenge –16		
Warm-Up –7			Warm-Up –17		
Practice –7			Practice –17		
Challenge –7			Challenge –17		
Warm-Up –8			Warm-Up –18		
Practice –8			Practice –18		
Challenge –8			Challenge –18		
Warm-Up –9			Warm-Up –19		
Practice –9			Practice –19		
Challenge –9			Challenge –19		
Warm-Up –10			Review 1–20A		
Practice –10			Review 1–20B		
Challenge –10			Addition and Subtraction 1–20A		
Warm-Up –11			Addition and Subtraction 1–20B		
Practice –11			Addition and Subtraction 1–20C		
Challenge –11			Addition and Subtraction 1–20D		

Timed Tests: Addition and Subtraction © 2004 Creative Teaching Press

Name _____   Date _____

Goal Time _____   Actual Time _____   Score _____

 # Subtraction Warm-Up   −2

1 7
   − 2

2 15
   − 2

3 6
   − 2

4 10
   − 2

5 17
   − 2

6 18
   − 2

7 22
   − 2

8 14
   − 2

9 5
   − 2

10 13
   − 2

11 9
   − 2

12 20
   − 2

13 3
   − 2

14 11
   − 2

15 16
   − 2

16 4
   − 2

17 12
   − 2

18 19
   − 2

19 8
   − 2

20 21
   − 2

Timed Tests: Addition and Subtraction © 2004 Creative Teaching Press

Name _____ Date _____

Goal Time _____ Actual Time _____ Score _____

 # Subtraction Practice   −2

1 3
−2

2 9
−2

3 20
−2

4 19
−2

5 14
−2

6 11
−2

7 4
−2

8 15
−2

9 12
−2

10 17
−2

11 21
−2

12 7
−2

13 5
−2

14 18
−2

15 10
−2

16 8
−2

17 13
−2

18 6
−2

19 22
−2

20 16
−2

Timed Tests: Addition and Subtraction © 2004 Creative Teaching Press

Subtraction Challenge −2

1 ☐
  −2
  7

2 ☐
  −2
  12

3 ☐
  −2
  3

4 ☐
  −2
  15

5 ☐
  −2
  10

6 ☐
  −2
  14

7 ☐
  −2
  1

8 ☐
  −2
  19

9 ☐
  −2
  8

10 ☐
  −2
  20

11 ☐
  −2
  5

12 ☐
  −2
  17

13 ☐
  −2
  11

14 ☐
  −2
  4

15 ☐
  −2
  6

16 ☐
  −2
  9

17 ☐
  −2
  16

18 ☐
  −2
  2

19 ☐
  −2
  18

20 ☐
  −2
  13

Timed Tests: Addition and Subtraction © 2004 Creative Teaching Press

Name _____   Date _____

 # Subtraction Warm-Up   −3

1   5
   − 3

2   14
   − 3

3   20
   − 3

4   7
   − 3

5   11
   − 3

6   17
   − 3

7   10
   − 3

8   23
   − 3

9   21
   − 3

10   18
   − 3

11   12
   − 3

12   6
   − 3

13   19
   − 3

14   15
   − 3

15   8
   − 3

16   9
   − 3

17   16
   − 3

18   22
   − 3

19   4
   − 3

20   13
   − 3

Timed Tests: Addition and Subtraction © 2004 Creative Teaching Press

Name _____ Date _____

Goal Time _____ Actual Time _____ Score _____

 # Subtraction Practice   −3

1 23
  − 3

2 12
  − 3

3 18
  − 3

4   7
  − 3

5 15
  − 3

6 22
  − 3

7 16
  − 3

8   4
  − 3

9 19
  − 3

10 10
  − 3

11   6
  − 3

12 20
  − 3

13   8
  − 3

14 13
  − 3

15   9
  − 3

16 14
  − 3

17 11
  − 3

18 17
  − 3

19 21
  − 3

20   5
  − 3

68

Timed Tests: Addition and Subtraction © 2004 Creative Teaching Press

Name _____ Date _____

 # Subtraction Challenge   −3

1 ☐
−3
—
1

2 ☐
−3
—
13

3 ☐
−3
—
6

4 ☐
−3
—
10

5 ☐
−3
—
17

6 ☐
−3
—
8

7 ☐
−3
—
15

8 ☐
−3
—
20

9 ☐
−3
—
5

10 ☐
−3
—
18

11 ☐
−3
—
11

12 ☐
−3
—
2

13 ☐
−3
—
19

14 ☐
−3
—
7

15 ☐
−3
—
14

16 ☐
−3
—
16

17 ☐
−3
—
4

18 ☐
−3
—
9

19 ☐
−3
—
12

20 ☐
−3
—
3

Timed Tests: Addition and Subtraction © 2004 Creative Teaching Press

Goal Time _____ Actual Time _____ Score _____

 # Subtraction Warm-Up −4

1 13
  − 4

2 15
  − 4

3 7
  − 4

4 19
  − 4

5 23
  − 4

6 6
  − 4

7 12
  − 4

8 22
  − 4

9 8
  − 4

10 11
  − 4

11 9
  − 4

12 20
  − 4

13 24
  − 4

14 17
  − 4

15 21
  − 4

16 18
  − 4

17 16
  − 4

18 10
  − 4

19 14
  − 4

20 5
  − 4

Timed Tests: Addition and Subtraction © 2004 Creative Teaching Press

Name _____ Date _____

Goal Time _____ Actual Time _____ Score _____

 # Subtraction Practice   −4

1   7
   − 4

2   10
   − 4

3   19
   − 4

4   13
   − 4

5   22
   − 4

6   15
   − 4

7   20
   − 4

8   5
   − 4

9   11
   − 4

10   18
   − 4

11   6
   − 4

12   14
   − 4

13   17
   − 4

14   9
   − 4

15   23
   − 4

16   12
   − 4

17   8
   − 4

18   21
   − 4

19   24
   − 4

20   16
   − 4

Timed Tests: Addition and Subtraction © 2004 Creative Teaching Press

Subtraction Challenge   −4

❶ ☐
  − 4
  13

❷ ☐
  − 4
  7

❸ ☐
  − 4
  1

❹ ☐
  − 4
  15

❺ ☐
  − 4
  4

❻ ☐
  − 4
  2

❼ ☐
  − 4
  9

❽ ☐
  − 4
  18

❾ ☐
  − 4
  6

❿ ☐
  − 4
  19

⓫ ☐
  − 4
  16

⓬ ☐
  − 4
  5

⓭ ☐
  − 4
  12

⓮ ☐
  − 4
  17

⓯ ☐
  − 4
  10

⓰ ☐
  − 4
  11

⓱ ☐
  − 4
  20

⓲ ☐
  − 4
  8

⓳ ☐
  − 4
  3

⓴ ☐
  − 4
  14

Timed Tests: Addition and Subtraction © 2004 Creative Teaching Press

Name _____ Date _____

Goal Time _____ Actual Time _____ Score _____

 # Subtraction Warm-Up   −5

1   19
       − 5

2   12
       − 5

3   23
       − 5

4   8
       − 5

5   16
       − 5

6   10
       − 5

7   7
       − 5

8   24
       − 5

9   14
       − 5

10   20
        − 5

11   17
        − 5

12   13
        − 5

13   21
        − 5

14   25
        − 5

15   11
        − 5

16   6
        − 5

17   22
        − 5

18   15
        − 5

19   18
        − 5

20   9
        − 5

Timed Tests: Addition and Subtraction © 2004 Creative Teaching Press

Name _____   Date _____

Goal Time _____   Actual Time _____   Score _____

 # Subtraction Practice   −5

1   15
     − 5

2   10
     − 5

3   23
     − 5

4   12
     − 5

5   8
     − 5

6   19
     − 5

7   13
     − 5

8   6
     − 5

9   20
     − 5

10   17
      − 5

11   7
      − 5

12   16
      − 5

13   22
      − 5

14   11
      − 5

15   24
      − 5

16   18
      − 5

17   21
      − 5

18   9
      − 5

19   14
      − 5

20   25
      − 5

Timed Tests: Addition and Subtraction © 2004 Creative Teaching Press

Name _____ Date _____

Goal Time _____ Actual Time _____ Score _____

 # Subtraction Challenge   −5

1 ☐
  − 5
  ‾‾‾
  13

2 ☐
  − 5
  ‾‾‾
  8

3 ☐
  − 5
  ‾‾‾
  3

4 ☐
  − 5
  ‾‾‾
  12

5 ☐
  − 5
  ‾‾‾
  17

6 ☐
  − 5
  ‾‾‾
  10

7 ☐
  − 5
  ‾‾‾
  6

8 ☐
  − 5
  ‾‾‾
  14

9 ☐
  − 5
  ‾‾‾
  7

10 ☐
  − 5
  ‾‾‾
  20

11 ☐
  − 5
  ‾‾‾
  5

12 ☐
  − 5
  ‾‾‾
  2

13 ☐
  − 5
  ‾‾‾
  9

14 ☐
  − 5
  ‾‾‾
  19

15 ☐
  − 5
  ‾‾‾
  11

16 ☐
  − 5
  ‾‾‾
  16

17 ☐
  − 5
  ‾‾‾
  15

18 ☐
  − 5
  ‾‾‾
  1

19 ☐
  − 5
  ‾‾‾
  4

20 ☐
  − 5
  ‾‾‾
  18

Timed Tests: Addition and Subtraction © 2004 Creative Teaching Press

Name _____ Date _____

Goal Time _____ Actual Time _____ Score _____

 # Subtraction Warm-Up   −6

1 9
   − 6

2 16
   − 6

3 25
   − 6

4 13
   − 6

5 21
   − 6

6 18
   − 6

7 11
   − 6

8 20
   − 6

9 17
   − 6

10 26
   − 6

11 14
   − 6

12 7
   − 6

13 22
   − 6

14 10
   − 6

15 24
   − 6

16 23
   − 6

17 12
   − 6

18 19
   − 6

19 8
   − 6

20 15
   − 6

Timed Tests: Addition and Subtraction © 2004 Creative Teaching Press

Name _____ Date _____

Goal Time _____ Actual Time _____ Score _____

 # Subtraction Practice   −6

1 17
− 6

2 12
− 6

3 20
− 6

4 10
− 6

5 22
− 6

6 14
− 6

7 9
− 6

8 25
− 6

9 15
− 6

10 7
− 6

11 21
− 6

12 26
− 6

13 24
− 6

14 13
− 6

15 18
− 6

16 19
− 6

17 11
− 6

18 16
− 6

19 8
− 6

20 23
− 6

Timed Tests: Addition and Subtraction © 2004 Creative Teaching Press

Subtraction Challenge   −6

1 ☐
− 6
15

2 ☐
− 6
4

3 ☐
− 6
17

4 ☐
− 6
9

5 ☐
− 6
19

6 ☐
− 6
8

7 ☐
− 6
11

8 ☐
− 6
5

9 ☐
− 6
2

10 ☐
− 6
14

11 ☐
− 6
13

12 ☐
− 6
1

13 ☐
− 6
20

14 ☐
− 6
6

15 ☐
− 6
12

16 ☐
− 6
7

17 ☐
− 6
16

18 ☐
− 6
3

19 ☐
− 6
10

20 ☐
− 6
18

Timed Tests: Addition and Subtraction © 2004 Creative Teaching Press

Name _____ Date _____

Goal Time _____ Actual Time _____ Score _____

 # Subtraction Warm-Up   −7

1 16
      − 7

2 22
      − 7

3 10
      − 7

4 25
      − 7

5 14
      − 7

6   8
      − 7

7 15
      − 7

8 19
      − 7

9 11
      − 7

10 24
       − 7

11 12
       − 7

12 23
       − 7

13 27
       − 7

14 18
       − 7

15 20
       − 7

16 21
       − 7

17 17
       − 7

18 26
       − 7

19 13
       − 7

20   9
       − 7

Timed Tests: Addition and Subtraction © 2004 Creative Teaching Press

Subtraction Practice   –7

1 27
− 7

2 20
− 7

3 25
− 7

4 12
− 7

5 23
− 7

6 18
− 7

7 11
− 7

8 22
− 7

9 15
− 7

10 19
− 7

11 24
− 7

12 13
− 7

13 26
− 7

14 8
− 7

15 17
− 7

16 16
− 7

17 9
− 7

18 14
− 7

19 21
− 7

20 10
− 7

Timed Tests: Addition and Subtraction © 2004 Creative Teaching Press

Name _____ Date _____

Goal Time _____ Actual Time _____ Score _____

 # Subtraction Challenge −7

◐ ☐
− 7
20

2 ☐
− 7
3

3 ☐
− 7
11

4 ☐
− 7
15

5 ☐
− 7
8

6 ☐
− 7
13

7 ☐
− 7
1

8 ☐
− 7
16

9 ☐
− 7
5

10 ☐
− 7
18

( ) ☐
− 7
6

12 ☐
− 7
19

13 ☐
− 7
9

14 ☐
− 7
2

15 ☐
− 7
14

16 ☐
− 7
17

17 ☐
− 7
4

18 ☐
− 7
10

19 ☐
− 7
7

20 ☐
− 7
12

Timed Tests: Addition and Subtraction © 2004 Creative Teaching Press

Name _____ Date _____

Goal Time _____ Actual Time _____ Score _____

 # Subtraction Warm-Up   −8

1 18
   − 8

2 11
   − 8

3 26
   − 8

4 15
   − 8

5 22
   − 8

6 13
   − 8

7 16
   − 8

8 24
   − 8

9 10
   − 8

10 19
   − 8

11 20
    − 8

12   9
    − 8

13 27
    − 8

14 21
    − 8

15 14
    − 8

16 23
    − 8

17 12
    − 8

18 25
    − 8

19 28
    − 8

20 17
    − 8

Timed Tests: Addition and Subtraction © 2004 Creative Teaching Press

Name _____ Date _____

Goal Time _____ Actual Time _____ Score _____

 # Subtraction Practice   −8

1 28
− 8

2 21
− 8

3 25
− 8

4 19
− 8

5 12
− 8

6 17
− 8

7 10
− 8

8 27
− 8

9 15
− 8

10 24
− 8

11 14
− 8

12 23
− 8

13 18
− 8

14 11
− 8

15 22
− 8

16 26
− 8

17 13
− 8

18 20
− 8

19 9
− 8

20 16
− 8

Timed Tests: Addition and Subtraction © 2004 Creative Teaching Press

Name _____ Date _____

Goal Time _____ Actual Time _____ Score _____

 # Subtraction Challenge −8

1
☐
− 8
16

2
☐
− 8
12

3
☐
− 8
4

4
☐
− 8
20

5
☐
− 8
8

6
☐
− 8
2

7
☐
− 8
10

8
☐
− 8
17

9
☐
− 8
6

10
☐
− 8
14

11
☐
− 8
7

12
☐
− 8
13

13
☐
− 8
1

14
☐
− 8
15

15
☐
− 8
18

16
☐
− 8
19

17
☐
− 8
5

18
☐
− 8
11

19
☐
− 8
9

20
☐
− 8
3

Timed Tests: Addition and Subtraction © 2004 Creative Teaching Press

Name _____ Date _____

Subtraction Warm-Up   −9

1 11
  − 9

2 23
  − 9

3 15
  − 9

4 26
  − 9

5 19
  − 9

6 28
  − 9

7 17
  − 9

8 27
  − 9

9 12
  − 9

10 25
  − 9

11 20
  − 9

12 14
  − 9

13 29
  − 9

14 24
  − 9

15 16
  − 9

16 22
  − 9

17 13
  − 9

18 21
  − 9

19 18
  − 9

20 10
  − 9

Timed Tests: Addition and Subtraction © 2004 Creative Teaching Press

Subtraction Practice −9

1 15
     − 9

2 26
     − 9

3 19
     − 9

4 27
     − 9

5 22
     − 9

6 23
     − 9

7 16
     − 9

8 11
     − 9

9 25
     − 9

10 28
      − 9

11 12
      − 9

12 20
      − 9

13 17
      − 9

14 29
      − 9

15 14
      − 9

16 24
      − 9

17 13
      − 9

18 21
      − 9

19 10
      − 9

20 18
      − 9

Timed Tests: Addition and Subtraction © 2004 Creative Teaching Press

Name _____ Date _____

Goal Time _____ Actual Time _____ Score _____

 # Subtraction Challenge   −9

1 ☐
−9
─────
16

2 ☐
−9
─────
7

3 ☐
−9
─────
13

4 ☐
−9
─────
5

5 ☐
−9
─────
10

6 ☐
−9
─────
3

7 ☐
−9
─────
11

8 ☐
−9
─────
18

9 ☐
−9
─────
2

10 ☐
−9
─────
19

11 ☐
−9
─────
6

12 ☐
−9
─────
15

13 ☐
−9
─────
1

14 ☐
−9
─────
12

15 ☐
−9
─────
8

16 ☐
−9
─────
14

17 ☐
−9
─────
9

18 ☐
−9
─────
20

19 ☐
−9
─────
4

20 ☐
−9
─────
17

Timed Tests: Addition and Subtraction © 2004 Creative Teaching Press

Name _____ Date _____

Goal Time _____ Actual Time _____ Score _____

 # Subtraction Warm-Up   −10

1   25
       −10

2   19
       −10

3   30
       −10

4   15
       −10

5   22
       −10

6   27
       −10

7   23
       −10

8   14
       −10

9   20
       −10

10   12
       −10

11   29
       −10

12   18
       −10

13   11
       −10

14   24
       −10

15   26
       −10

16   16
       −10

17   21
       −10

18   17
       −10

19   28
       −10

20   13
       −10

Timed Tests: Addition and Subtraction © 2004 Creative Teaching Press

Name _____ Date _____

Goal Time _____ Actual Time _____ Score _____

 # Subtraction Practice −10

1 $\begin{array}{r} 21 \\ -10 \\ \hline \end{array}$ **2** $\begin{array}{r} 14 \\ -10 \\ \hline \end{array}$ **3** $\begin{array}{r} 20 \\ -10 \\ \hline \end{array}$ **4** $\begin{array}{r} 17 \\ -10 \\ \hline \end{array}$ **5** $\begin{array}{r} 22 \\ -10 \\ \hline \end{array}$

6 $\begin{array}{r} 13 \\ -10 \\ \hline \end{array}$ **7** $\begin{array}{r} 25 \\ -10 \\ \hline \end{array}$ **8** $\begin{array}{r} 29 \\ -10 \\ \hline \end{array}$ **9** $\begin{array}{r} 15 \\ -10 \\ \hline \end{array}$ **10** $\begin{array}{r} 26 \\ -10 \\ \hline \end{array}$

11 $\begin{array}{r} 18 \\ -10 \\ \hline \end{array}$ **12** $\begin{array}{r} 27 \\ -10 \\ \hline \end{array}$ **13** $\begin{array}{r} 12 \\ -10 \\ \hline \end{array}$ **14** $\begin{array}{r} 24 \\ -10 \\ \hline \end{array}$ **15** $\begin{array}{r} 19 \\ -10 \\ \hline \end{array}$

16 $\begin{array}{r} 23 \\ -10 \\ \hline \end{array}$ **17** $\begin{array}{r} 16 \\ -10 \\ \hline \end{array}$ **18** $\begin{array}{r} 30 \\ -10 \\ \hline \end{array}$ **19** $\begin{array}{r} 28 \\ -10 \\ \hline \end{array}$ **20** $\begin{array}{r} 11 \\ -10 \\ \hline \end{array}$

Timed Tests: Addition and Subtraction © 2004 Creative Teaching Press

Name _____ Date _____

Goal Time _____ Actual Time _____ Score _____

 # Subtraction Challenge −10

1 ☐
−10
12

2 ☐
−10
9

3 ☐
−10
17

4 ☐
−10
5

5 ☐
−10
19

6 ☐
−10
4

7 ☐
−10
16

8 ☐
−10
7

9 ☐
−10
2

10 ☐
−10
11

11 ☐
−10
18

12 ☐
−10
1

13 ☐
−10
10

14 ☐
−10
8

15 ☐
−10
14

16 ☐
−10
15

17 ☐
−10
6

18 ☐
−10
13

19 ☐
−10
20

20 ☐
−10
3

Timed Tests: Addition and Subtraction © 2004 Creative Teaching Press

Name _____ Date _____

Goal Time _____ Actual Time _____ Score _____

 # Subtraction Warm-Up −11

1 23
−11

2 18
−11

3 24
−11

4 30
−11

5 16
−11

6 27
−11

7 15
−11

8 12
−11

9 21
−11

10 28
−11

11 17
−11

12 26
−11

13 20
−11

14 29
−11

15 14
−11

16 13
−11

17 22
−11

18 31
−11

19 19
−11

20 25
−11

Timed Tests: Addition and Subtraction © 2004 Creative Teaching Press

Name _____ Date _____

Goal Time _____ Actual Time _____ Score _____

 # Subtraction Practice   −11

1 30
−11

2 23
−11

3 16
−11

4 25
−11

5 19
−11

6 12
−11

7 18
−11

8 31
−11

9 15
−11

10 26
−11

11 27
−11

12 21
−11

13 20
−11

14 24
−11

15 13
−11

16 29
−11

17 28
−11

18 17
−11

19 14
−11

20 22
−11

Timed Tests: Addition and Subtraction © 2004 Creative Teaching Press

Name _____ Date _____

Goal Time _____ Actual Time _____ Score _____

 # Subtraction Challenge   −11

1 ☐
−11
___
1

2 ☐
−11
___
15

3 ☐
−11
___
10

4 ☐
−11
___
4

5 ☐
−11
___
13

6 ☐
−11
___
17

7 ☐
−11
___
3

8 ☐
−11
___
12

9 ☐
−11
___
7

10 ☐
−11
___
19

11 ☐
−11
___
5

12 ☐
−11
___
16

13 ☐
−11
___
20

14 ☐
−11
___
8

15 ☐
−11
___
2

16 ☐
−11
___
14

17 ☐
−11
___
11

18 ☐
−11
___
6

19 ☐
−11
___
18

20 ☐
−11
___
9

Timed Tests: Addition and Subtraction © 2004 Creative Teaching Press

Name _____ Date _____

Goal Time _____ Actual Time _____ Score _____

 # Subtraction Warm-Up −12

1 16
     −12

2 30
     −12

3 28
     −12

4 23
     −12

5 15
     −12

6 22
     −12

7 14
     −12

8 32
     −12

9 17
     −12

10 25
      −12

11 19
      −12

12 26
      −12

13 21
      −12

14 13
      −12

15 29
      −12

16 24
      −12

17 31
      −12

18 18
      −12

19 20
      −12

20 27
      −12

Timed Tests: Addition and Subtraction © 2004 Creative Teaching Press

Name _____ Date _____

Goal Time _____ Actual Time _____ Score _____

 **Subtraction Practice   −12**

1   32
     −12

2   18
     −12

3   24
     −12

4   16
     −12

5   27
     −12

6   15
     −12

7   21
     −12

8   28
     −12

9   17
     −12

10   22
      −12

11   30
      −12

12   14
      −12

13   29
      −12

14   23
      −12

15   20
      −12

16   19
      −12

17   25
      −12

18   26
      −12

19   13
      −12

20   31
      −12

Name _____   Date _____

Goal Time _____   Actual Time _____   Score _____

 **Subtraction Challenge** −12

1 ☐
−12
  9

2 ☐
−12
 14

3 ☐
−12
  3

4 ☐
−12
 11

5 ☐
−12
 19

6 ☐
−12
  5

7 ☐
−12
 16

8 ☐
−12
  7

9 ☐
−12
 13

10 ☐
−12
  1

11 ☐
−12
 18

12 ☐
−12
  2

13 ☐
−12
 10

14 ☐
−12
  6

15 ☐
−12
 17

16 ☐
−12
 12

17 ☐
−12
  8

18 ☐
−12
 20

19 ☐
−12
  4

20 ☐
−12
 15

Timed Tests: Addition and Subtraction © 2004 Creative Teaching Press

Name _____ Date _____

Goal Time _____ Actual Time _____ Score _____

 # Subtraction Warm-Up −13

1 21
−13

2 14
−13

3 30
−13

4 25
−13

5 18
−13

6 17
−13

7 29
−13

8 22
−13

9 32
−13

10 27
−13

11 24
−13

12 33
−13

13 15
−13

14 31
−13

15 20
−13

16 26
−13

17 19
−13

18 23
−13

19 28
−13

20 16
−13

Timed Tests: Addition and Subtraction © 2004 Creative Teaching Press

Name _____ Date _____

Goal Time _____ Actual Time _____ Score _____

 # Subtraction Practice −13

1 28
    −13

2 30
    −13

3 25
    −13

4 31
    −13

5 33
    −13

6 16
    −13

7 26
    −13

8 19
    −13

9 29
    −13

10 32
    −13

11 27
    −13

12 24
    −13

13 15
    −13

14 20
    −13

15 21
    −13

16 14
    −13

17 18
    −13

18 23
    −13

19 17
    −13

20 22
    −13

Timed Tests: Addition and Subtraction © 2004 Creative Teaching Press

Name _____ Date _____

Goal Time _____ Actual Time _____ Score _____

 # Subtraction Challenge   −13

1 ☐
−13
5

2 ☐
−13
17

3 ☐
−13
9

4 ☐
−13
3

5 ☐
−13
11

6 ☐
−13
14

7 ☐
−13
19

8 ☐
−13
1

9 ☐
−13
18

10 ☐
−13
7

11 ☐
−13
8

12 ☐
−13
16

13 ☐
−13
4

14 ☐
−13
13

15 ☐
−13
12

16 ☐
−13
2

17 ☐
−13
20

18 ☐
−13
10

19 ☐
−13
6

20 ☐
−13
15

Timed Tests: Addition and Subtraction © 2004 Creative Teaching Press

Name _____   Date _____

Goal Time _____   Actual Time _____   Score _____

 # Subtraction Warm-Up −14

① 17
−14

② 23
−14

③ 29
−14

④ 20
−14

⑤ 26
−14

⑥ 25
−14

⑦ 19
−14

⑧ 33
−14

⑨ 15
−14

⑩ 31
−14

⑪ 28
−14

⑫ 34
−14

⑬ 22
−14

⑭ 30
−14

⑮ 18
−14

⑯ 21
−14

⑰ 32
−14

⑱ 16
−14

⑲ 27
−14

⑳ 24
−14

Timed Tests: Addition and Subtraction © 2004 Creative Teaching Press

Name _____ Date _____

Goal Time _____ Actual Time _____ Score _____

 # Subtraction Practice   −14

1 30
−14

2 27
−14

3 29
−14

4 20
−14

5 26
−14

6 23
−14

7 18
−14

8 16
−14

9 15
−14

10 17
−14

11 19
−14

12 28
−14

13 34
−14

14 22
−14

15 32
−14

16 25
−14

17 21
−14

18 24
−14

19 33
−14

20 31
−14

Timed Tests: Addition and Subtraction © 2004 Creative Teaching Press

Name _____ Date _____

Goal Time _____ Actual Time _____ Score _____

 # Subtraction Challenge   −14

1 ☐
−14
2

2 ☐
−14
9

3 ☐
−14
15

4 ☐
−14
19

5 ☐
−14
6

6 ☐
−14
13

7 ☐
−14
5

8 ☐
−14
11

9 ☐
−14
18

10 ☐
−14
3

11 ☐
−14
7

12 ☐
−14
14

13 ☐
−14
1

14 ☐
−14
10

15 ☐
−14
20

16 ☐
−14
4

17 ☐
−14
17

18 ☐
−14
12

19 ☐
−14
8

20 ☐
−14
16

Timed Tests: Addition and Subtraction © 2004 Creative Teaching Press

Name _____ Date _____

Goal Time _____ Actual Time _____ Score _____

 # Subtraction Warm-Up −15

1 26
     −15

2 21
     −15

3 29
     −15

4 18
     −15

5 24
     −15

6 17
     −15

7 32
     −15

8 20
     −15

9 35
     −15

10 28
      −15

11 33
      −15

12 19
      −15

13 27
      −15

14 23
      −15

15 30
      −15

16 22
      −15

17 31
      −15

18 25
      −15

19 34
      −15

20 16
      −15

Timed Tests: Addition and Subtraction © 2004 Creative Teaching Press

 # Subtraction Practice   −15

❶ 23
   −15

❷ 20
   −15

❸ 30
   −15

❹ 33
   −15

❺ 22
   −15

❻ 31
   −15

❼ 17
   −15

❽ 32
   −15

❾ 19
   −15

❿ 28
   −15

⓫ 26
   −15

⓬ 24
   −15

⓭ 29
   −15

⓮ 34
   −15

⓯ 27
   −15

⓰ 21
   −15

⓱ 18
   −15

⓲ 35
   −15

⓳ 25
   −15

⓴ 16
   −15

Timed Tests: Addition and Subtraction © 2004 Creative Teaching Press

Name _____ Date _____

Goal Time _____ Actual Time _____ Score _____

 # Subtraction Challenge   −15

1) ☐
   −15
   ‾‾1‾

2 ☐
   −15
   ‾‾8‾

3 ☐
   −15
   ‾12‾

4 ☐
   −15
   ‾‾6‾

5 ☐
   −15
   ‾18‾

6 ☐
   −15
   ‾‾5‾

7 ☐
   −15
   ‾10‾

8 ☐
   −15
   ‾16‾

9 ☐
   −15
   ‾‾3‾

10 ☐
   −15
   ‾14‾

11 ☐
   −15
   ‾13‾

12 ☐
   −15
   ‾‾2‾

13 ☐
   −15
   ‾20‾

14 ☐
   −15
   ‾‾9‾

15 ☐
   −15
   ‾17‾

16 ☐
   −15
   ‾19‾

17 ☐
   −15
   ‾‾7‾

18 ☐
   −15
   ‾15‾

19 ☐
   −15
   ‾‾4‾

20 ☐
   −15
   ‾11‾

Timed Tests: Addition and Subtraction © 2004 Creative Teaching Press

Name _____   Date _____

Goal Time _____   Actual Time _____   Score _____

 # Subtraction Warm-Up   −16

1 20
     −16

2 28
     −16

3 36
     −16

4 22
     −16

5 34
     −16

6 24
     −16

7 17
     −16

8 32
     −16

9 30
     −16

10 26
      −16

11 33
      −16

12 21
      −16

13 29
      −16

14 18
      −16

15 35
      −16

16 27
      −16

17 19
      −16

18 23
      −16

19 31
      −16

20 25
      −16

Timed Tests: Addition and Subtraction © 2004 Creative Teaching Press

Name _____   Date _____

Goal Time _____   Actual Time _____   Score _____

 # Subtraction Practice   −16

1
$$\begin{array}{r} 21 \\ -16 \\ \hline \end{array}$$

2
$$\begin{array}{r} 28 \\ -16 \\ \hline \end{array}$$

3
$$\begin{array}{r} 32 \\ -16 \\ \hline \end{array}$$

4
$$\begin{array}{r} 26 \\ -16 \\ \hline \end{array}$$

5
$$\begin{array}{r} 18 \\ -16 \\ \hline \end{array}$$

6
$$\begin{array}{r} 17 \\ -16 \\ \hline \end{array}$$

7
$$\begin{array}{r} 30 \\ -16 \\ \hline \end{array}$$

8
$$\begin{array}{r} 23 \\ -16 \\ \hline \end{array}$$

9
$$\begin{array}{r} 34 \\ -16 \\ \hline \end{array}$$

10
$$\begin{array}{r} 36 \\ -16 \\ \hline \end{array}$$

11
$$\begin{array}{r} 25 \\ -16 \\ \hline \end{array}$$

12
$$\begin{array}{r} 35 \\ -16 \\ \hline \end{array}$$

13
$$\begin{array}{r} 19 \\ -16 \\ \hline \end{array}$$

14
$$\begin{array}{r} 33 \\ -16 \\ \hline \end{array}$$

15
$$\begin{array}{r} 24 \\ -16 \\ \hline \end{array}$$

16
$$\begin{array}{r} 22 \\ -16 \\ \hline \end{array}$$

17
$$\begin{array}{r} 27 \\ -16 \\ \hline \end{array}$$

18
$$\begin{array}{r} 31 \\ -16 \\ \hline \end{array}$$

19
$$\begin{array}{r} 20 \\ -16 \\ \hline \end{array}$$

20
$$\begin{array}{r} 29 \\ -16 \\ \hline \end{array}$$

Timed Tests: Addition and Subtraction © 2004 Creative Teaching Press

Name _____ Date _____

Goal Time _____ Actual Time _____ Score _____

 # Subtraction Challenge   −16

1 ☐
  −16
  4

2 ☐
  −16
  13

3 ☐
  −16
  7

4 ☐
  −16
  15

5 ☐
  −16
  1

6 ☐
  −16
  9

7 ☐
  −16
  17

8 ☐
  −16
  5

9 ☐
  −16
  11

10 ☐
  −16
  20

11 ☐
  −16
  14

12 ☐
  −16
  2

13 ☐
  −16
  8

14 ☐
  −16
  16

15 ☐
  −16
  19

16 ☐
  −16
  18

17 ☐
  −16
  6

18 ☐
  −16
  10

19 ☐
  −16
  3

20 ☐
  −16
  12

Timed Tests: Addition and Subtraction © 2004 Creative Teaching Press

Name _____ Date _____

Goal Time _____ Actual Time _____ Score _____

 # Subtraction Warm-Up −17

1 29
−17

2 18
−17

3 35
−17

4 23
−17

5 27
−17

6 25
−17

7 33
−17

8 21
−17

9 37
−17

10 31
−17

11 22
−17

12 34
−17

13 30
−17

14 19
−17

15 28
−17

16 20
−17

17 26
−17

18 32
−17

19 36
−17

20 24
−17

Timed Tests: Addition and Subtraction © 2004 Creative Teaching Press

Name _____ Date _____

Goal Time _____ Actual Time _____ Score _____

 # Subtraction Practice −17

1 22
    −17

2 30
    −17

3 27
    −17

4 34
    −17

5 20
    −17

6 19
    −17

7 33
    −17

8 24
    −17

9 37
    −17

10 31
    −17

11 36
    −17

12 23
    −17

13 29
    −17

14 18
    −17

15 26
    −17

16 25
    −17

17 32
    −17

18 21
    −17

19 28
    −17

20 35
    −17

Timed Tests: Addition and Subtraction © 2004 Creative Teaching Press

Name _____ Date _____

Goal Time _____ Actual Time _____ Score _____

 # Subtraction Challenge −17

1 ☐
−17
7

2 ☐
−17
13

3 ☐
−17
5

4 ☐
−17
16

5 ☐
−17
10

6 ☐
−17
4

7 ☐
−17
8

8 ☐
−17
17

9 ☐
−17
3

10 ☐
−17
14

11 ☐
−17
20

12 ☐
−17
11

13 ☐
−17
2

14 ☐
−17
15

15 ☐
−17
19

16 ☐
−17
1

17 ☐
−17
9

18 ☐
−17
18

19 ☐
−17
6

20 ☐
−17
12

Timed Tests: Addition and Subtraction © 2004 Creative Teaching Press

Name _____   Date _____

Goal Time _____   Actual Time _____   Score _____

 # Subtraction Warm-Up   −18

1 22
      −18

2 30
      −18

3 26
      −18

4 34
      −18

5 27
      −18

6 33
      −18

7 25
      −18

8 38
      −18

9 21
      −18

10 29
       −18

11 37
       −18

12 19
       −18

13 36
       −18

14 31
       −18

15 24
       −18

16 23
       −18

17 28
       −18

18 35
       −18

19 20
       −18

20 32
       −18

Timed Tests: Addition and Subtraction © 2004 Creative Teaching Press

Name _____ Date _____

Goal Time _____ Actual Time _____ Score _____

 # Subtraction Practice   −18

1 $\begin{array}{r} 19 \\ -18 \\ \hline \end{array}$ **2** $\begin{array}{r} 37 \\ -18 \\ \hline \end{array}$ **3** $\begin{array}{r} 22 \\ -18 \\ \hline \end{array}$ **4** $\begin{array}{r} 28 \\ -18 \\ \hline \end{array}$ **5** $\begin{array}{r} 31 \\ -18 \\ \hline \end{array}$

6 $\begin{array}{r} 30 \\ -18 \\ \hline \end{array}$ **7** $\begin{array}{r} 26 \\ -18 \\ \hline \end{array}$ **8** $\begin{array}{r} 36 \\ -18 \\ \hline \end{array}$ **9** $\begin{array}{r} 24 \\ -18 \\ \hline \end{array}$ **10** $\begin{array}{r} 34 \\ -18 \\ \hline \end{array}$

11 $\begin{array}{r} 32 \\ -18 \\ \hline \end{array}$ **12** $\begin{array}{r} 21 \\ -18 \\ \hline \end{array}$ **13** $\begin{array}{r} 38 \\ -18 \\ \hline \end{array}$ **14** $\begin{array}{r} 29 \\ -18 \\ \hline \end{array}$ **15** $\begin{array}{r} 25 \\ -18 \\ \hline \end{array}$

16 $\begin{array}{r} 23 \\ -18 \\ \hline \end{array}$ **17** $\begin{array}{r} 35 \\ -18 \\ \hline \end{array}$ **18** $\begin{array}{r} 27 \\ -18 \\ \hline \end{array}$ **19** $\begin{array}{r} 33 \\ -18 \\ \hline \end{array}$ **20** $\begin{array}{r} 20 \\ -18 \\ \hline \end{array}$

Timed Tests: Addition and Subtraction © 2004 Creative Teaching Press

Name _____ Date _____

Goal Time _____ Actual Time _____ Score _____

 # Subtraction Challenge   −18

1 ☐
−18
11

2 ☐
−18
4

3 ☐
−18
19

4 ☐
−18
7

5 ☐
−18
13

6 ☐
−18
5

7 ☐
−18
8

8 ☐
−18
14

9 ☐
−18
17

10 ☐
−18
2

11 ☐
−18
16

12 ☐
−18
12

13 ☐
−18
3

14 ☐
−18
20

15 ☐
−18
9

16 ☐
−18
18

17 ☐
−18
1

18 ☐
−18
10

19 ☐
−18
15

20 ☐
−18
6

114

Timed Tests: Addition and Subtraction © 2004 Creative Teaching Press

Name _____ Date _____

Goal Time _____ Actual Time _____ Score _____

 # Subtraction Warm-Up −19

1 23
     −19

2 30
     −19

3 37
     −19

4 24
     −19

5 35
     −19

6 27
     −19

7 39
     −19

8 20
     −19

9 33
     −19

10 26
      −19

11 21
      −19

12 34
      −19

13 28
      −19

14 31
      −19

15 38
      −19

16 32
      −19

17 25
      −19

18 36
      −19

19 22
      −19

20 29
      −19

Timed Tests: Addition and Subtraction © 2004 Creative Teaching Press

Subtraction Practice   −19

1 32
      −19

2 24
      −19

3 35
      −19

4 20
      −19

5 39
      −19

6 28
      −19

7 37
      −19

8 22
      −19

9 33
      −19

10 25
       −19

11 21
       −19

12 30
       −19

13 36
       −19

14 29
       −19

15 34
       −19

16 27
       −19

17 31
       −19

18 38
       −19

19 23
       −19

20 26
       −19

Timed Tests: Addition and Subtraction © 2004 Creative Teaching Press

Name _____ Date _____

Goal Time _____ Actual Time _____ Score _____

 **Subtraction Challenge −19**

1 ☐
−19
  9

2 ☐
−19
 17

3 ☐
−19
 12

4 ☐
−19
  6

5 ☐
−19
 20

6 ☐
−19
 11

7 ☐
−19
  4

8 ☐
−19
 16

9 ☐
−19
  2

10 ☐
−19
 14

11 ☐
−19
  8

12 ☐
−19
 15

13 ☐
−19
  1

14 ☐
−19
 10

15 ☐
−19
  7

16 ☐
−19
  3

17 ☐
−19
 19

18 ☐
−19
 13

19 ☐
−19
  5

20 ☐
−19
 18

Timed Tests: Addition and Subtraction © 2004 Creative Teaching Press

 # Subtraction Review   1–20A

1 $\begin{array}{r} 10 \\ -\ 7 \\ \hline \end{array}$
2 $\begin{array}{r} 19 \\ -12 \\ \hline \end{array}$
3 $\begin{array}{r} 14 \\ -\ 9 \\ \hline \end{array}$
4 $\begin{array}{r} 9 \\ -\ 4 \\ \hline \end{array}$
5 $\begin{array}{r} 7 \\ -\ 3 \\ \hline \end{array}$

6 $\begin{array}{r} 17 \\ -15 \\ \hline \end{array}$
7 $\begin{array}{r} 25 \\ -11 \\ \hline \end{array}$
8 $\begin{array}{r} 3 \\ -\ 2 \\ \hline \end{array}$
9 $\begin{array}{r} 20 \\ -16 \\ \hline \end{array}$
10 $\begin{array}{r} 18 \\ -\ 5 \\ \hline \end{array}$

11 $\begin{array}{r} 23 \\ -11 \\ \hline \end{array}$
12 $\begin{array}{r} 12 \\ -\ 6 \\ \hline \end{array}$
13 $\begin{array}{r} 26 \\ -19 \\ \hline \end{array}$
14 $\begin{array}{r} 21 \\ -13 \\ \hline \end{array}$
15 $\begin{array}{r} 25 \\ -14 \\ \hline \end{array}$

16 $\begin{array}{r} 13 \\ -\ 8 \\ \hline \end{array}$
17 $\begin{array}{r} 22 \\ -10 \\ \hline \end{array}$
18 $\begin{array}{r} 8 \\ -\ 7 \\ \hline \end{array}$
19 $\begin{array}{r} 29 \\ -17 \\ \hline \end{array}$
20 $\begin{array}{r} 30 \\ -18 \\ \hline \end{array}$

Timed Tests: Addition and Subtraction © 2004 Creative Teaching Press

Goal Time _____ Actual Time _____ Score _____

 **Subtraction Review   1–20B**

1 15
     −12

2 24
     −19

3 19
     −18

4 6
     − 3

5 12
     − 8

6 14
     − 7

7 20
     −13

8 22
     −17

9 4
     − 2

10 17
      − 9

11 11
      − 6

12 9
      − 4

13 13
      − 5

14 29
      −16

15 16
      −10

16 27
      −14

17 23
      −11

18 25
      −15

19 7
      − 2

20 14
      − 5

Timed Tests: Addition and Subtraction © 2004 Creative Teaching Press

Name _____   Date _____

Goal Time _____   Actual Time _____   Score _____

 # Addition and Subtraction
Review   1–20A

1   5
   + 9

2   31
   − 13

3   11
   + 3

4   9
   − 6

5   13
   − 8

6   16
   − 10

7   8
   + 8

8   13
   + 7

9   24
   − 14

10   18
   + 19

11   7
   − 5

12   12
   + 2

13   18
   − 7

14   15
   + 17

15   22
   − 16

16   26
   − 18

17   14
   + 4

18   17
   + 5

19   10
   − 4

20   10
   + 6

Timed Tests: Addition and Subtraction © 2004 Creative Teaching Press

Name _____   Date _____

Goal Time _____   Actual Time _____   Score _____

Addition and Subtraction
Review   1–20B

1 5
  + 5

2 23
  –11

3   7
  + 9

4   6
  +18

5   9
  – 2

6 28
  –15

7 13
  +15

8 12
  – 9

9   4
  + 8

10 25
  –16

11 14
  – 6

12 19
  +19

13 26
  –13

14 16
  + 9

15   2
  +16

16 32
  –19

17 11
  +11

18 13
  – 7

19 10
  +14

20 36
  –14

Timed Tests: Addition and Subtraction © 2004 Creative Teaching Press

Addition and Subtraction
Review   1–20C

1   22
      −12

2   4
      +17

3   15
      − 9

4   20
      − 8

5   12
      +13

6   19
      + 2

7   24
      −10

8   10
      +10

9   26
      −14

10 14
      +19

11   9
      − 5

12 20
      +16

13 35
      −18

14   5
      + 7

15 19
      −13

16 17
      +17

17 11
      − 7

18 15
      + 8

19 18
      + 3

20 30
      −16

Timed Tests: Addition and Subtraction © 2004 Creative Teaching Press

Name _____ Date _____

Goal Time _____ Actual Time _____ Score _____

 # Addition and Subtraction
Review   1–20D

1 $\begin{array}{r} 11 \\ +13 \\ \hline \end{array}$ **2** $\begin{array}{r} 8 \\ -4 \\ \hline \end{array}$ **3** $\begin{array}{r} 12 \\ +20 \\ \hline \end{array}$ **4** $\begin{array}{r} 30 \\ -17 \\ \hline \end{array}$ **5** $\begin{array}{r} 4 \\ +19 \\ \hline \end{array}$

6 $\begin{array}{r} 28 \\ -15 \\ \hline \end{array}$ **7** $\begin{array}{r} 18 \\ +18 \\ \hline \end{array}$ **8** $\begin{array}{r} 15 \\ -9 \\ \hline \end{array}$ **9** $\begin{array}{r} 3 \\ +5 \\ \hline \end{array}$ **10** $\begin{array}{r} 24 \\ -12 \\ \hline \end{array}$

11 $\begin{array}{r} 7 \\ +8 \\ \hline \end{array}$ **12** $\begin{array}{r} 10 \\ -6 \\ \hline \end{array}$ **13** $\begin{array}{r} 6 \\ +14 \\ \hline \end{array}$ **14** $\begin{array}{r} 27 \\ -11 \\ \hline \end{array}$ **15** $\begin{array}{r} 10 \\ +9 \\ \hline \end{array}$

16 $\begin{array}{r} 36 \\ -19 \\ \hline \end{array}$ **17** $\begin{array}{r} 12 \\ +12 \\ \hline \end{array}$ **18** $\begin{array}{r} 11 \\ -8 \\ \hline \end{array}$ **19** $\begin{array}{r} 16 \\ +17 \\ \hline \end{array}$ **20** $\begin{array}{r} 32 \\ -18 \\ \hline \end{array}$

Answer Key

Warm-Up +2
1. 10	2. 5	3. 8	4. 6	5. 16
6. 12	7. 3	8. 4	9. 21	10. 7
11. 15	12. 11	13. 17	14. 13	15. 19
16. 14	17. 20	18. 9	19. 22	20. 18

Warm-Up +5
1. 6	2. 20	3. 16	4. 12	5. 23
6. 22	7. 8	8. 24	9. 18	10. 13
11. 19	12. 14	13. 10	14. 21	15. 7
16. 9	17. 25	18. 17	19. 11	20. 15

Warm-Up +8
1. 15	2. 20	3. 17	4. 24	5. 9
6. 27	7. 10	8. 22	9. 14	10. 25
11. 12	12. 18	13. 26	14. 11	15. 23
16. 21	17. 13	18. 28	19. 16	20. 19

Practice +2
1. 7	2. 12	3. 10	4. 21	5. 17
6. 5	7. 3	8. 16	9. 14	10. 20
11. 8	12. 19	13. 11	14. 4	15. 13
16. 18	17. 6	18. 22	19. 15	20. 9

Practice +5
1. 19	2. 7	3. 25	4. 13	5. 21
6. 23	7. 9	8. 18	9. 11	10. 15
11. 16	12. 12	13. 20	14. 6	15. 17
16. 24	17. 8	18. 14	19. 22	20. 10

Practice +8
1. 10	2. 24	3. 16	4. 12	5. 27
6. 15	7. 19	8. 13	9. 22	10. 26
11. 9	12. 17	13. 28	14. 20	15. 14
16. 21	17. 25	18. 11	19. 23	20. 18

Challenge +2
1. 10	2. 3	3. 16	4. 9	5. 14
6. 6	7. 17	8. 1	9. 12	10. 19
11. 7	12. 5	13. 11	14. 8	15. 2
16. 15	17. 18	18. 4	19. 13	20. 20

Challenge +5
1. 5	2. 16	3. 3	4. 19	5. 10
6. 13	7. 18	8. 7	9. 20	10. 1
11. 11	12. 2	13. 14	14. 17	15. 8
16. 4	17. 9	18. 15	19. 6	20. 12

Challenge +8
1. 8	2. 15	3. 2	4. 18	5. 7
6. 10	7. 6	8. 19	9. 4	10. 11
11. 13	12. 1	13. 16	14. 9	15. 20
16. 17	17. 5	18. 12	19. 3	20. 14

Warm-Up +3
1. 6	2. 16	3. 9	4. 4	5. 17
6. 13	7. 10	8. 7	9. 22	10. 12
11. 8	12. 11	13. 19	14. 14	15. 20
16. 15	17. 23	18. 21	19. 5	20. 18

Warm-Up +6
1. 15	2. 24	3. 8	4. 20	5. 14
6. 26	7. 7	8. 22	9. 11	10. 18
11. 10	12. 19	13. 13	14. 25	15. 16
16. 23	17. 12	18. 17	19. 21	20. 9

Warm-Up +9
1. 19	2. 22	3. 13	4. 27	5. 15
6. 16	7. 25	8. 10	9. 18	10. 23
11. 11	12. 28	13. 20	14. 14	15. 26
16. 17	17. 24	18. 12	19. 29	20. 21

Practice +3
1. 12	2. 13	3. 6	4. 22	5. 19
6. 17	7. 18	8. 16	9. 15	10. 21
11. 11	12. 8	13. 20	14. 4	15. 9
16. 14	17. 5	18. 23	19. 10	20. 7

Practice +6
1. 20	2. 24	3. 9	4. 16	5. 12
6. 7	7. 18	8. 11	9. 14	10. 23
11. 22	12. 25	13. 8	14. 13	15. 17
16. 10	17. 21	18. 26	19. 15	20. 19

Practice +9
1. 29	2. 17	3. 25	4. 12	5. 28
6. 21	7. 23	8. 14	9. 19	10. 26
11. 27	12. 11	13. 18	14. 15	15. 22
16. 13	17. 20	18. 16	19. 24	20. 10

Challenge +3
1. 15	2. 7	3. 19	4. 4	5. 11
6. 9	7. 16	8. 1	9. 13	10. 18
11. 3	12. 10	13. 20	14. 6	15. 14
16. 12	17. 5	18. 8	19. 17	20. 2

Challenge +6
1. 1	2. 12	3. 6	4. 20	5. 9
6. 7	7. 15	8. 11	9. 4	10. 17
11. 18	12. 3	13. 16	14. 10	15. 2
16. 5	17. 13	18. 19	19. 8	20. 14

Challenge +9
1. 5	2. 13	3. 11	4. 2	5. 18
6. 7	7. 15	8. 3	9. 20	10. 9
11. 16	12. 1	13. 12	14. 6	15. 14
16. 10	17. 8	18. 17	19. 4	20. 19

Warm-Up +4
1. 19	2. 7	3. 16	4. 23	5. 10
6. 6	7. 21	8. 14	9. 12	10. 18
11. 17	12. 5	13. 9	14. 20	15. 13
16. 24	17. 11	18. 15	19. 22	20. 8

Warm-Up +7
1. 9	2. 17	3. 26	4. 12	5. 20
6. 21	7. 13	8. 8	9. 24	10. 15
11. 14	12. 25	13. 16	14. 18	15. 10
16. 23	17. 11	18. 19	19. 27	20. 22

Warm-Up +10
1. 13	2. 27	3. 18	4. 30	5. 23
6. 16	7. 14	8. 28	9. 22	10. 17
11. 24	12. 20	13. 12	14. 15	15. 29
16. 11	17. 21	18. 19	19. 26	20. 25

Practice +4
1. 18	2. 6	3. 23	4. 12	5. 13
6. 8	7. 22	8. 15	9. 9	10. 19
11. 14	12. 10	13. 21	14. 16	15. 5
16. 17	17. 24	18. 20	19. 7	20. 11

Practice +7
1. 16	2. 20	3. 12	4. 23	5. 18
6. 13	7. 24	8. 15	9. 9	10. 27
11. 25	12. 10	13. 17	14. 22	15. 11
16. 8	17. 21	18. 14	19. 26	20. 19

Practice +10
1. 19	2. 17	3. 22	4. 28	5. 13
6. 14	7. 29	8. 11	9. 25	10. 23
11. 24	12. 12	13. 21	14. 18	15. 26
16. 15	17. 20	18. 16	19. 27	20. 30

Challenge +4
1. 3	2. 17	3. 11	4. 7	5. 15
6. 12	7. 9	8. 20	9. 5	10. 18
11. 6	12. 14	13. 1	14. 16	15. 10
16. 13	17. 4	18. 19	19. 8	20. 2

Challenge +7
1. 15	2. 6	3. 18	4. 11	5. 3
6. 9	7. 1	8. 13	9. 5	10. 19
11. 17	12. 10	13. 7	14. 2	15. 14
16. 4	17. 20	18. 12	19. 16	20. 8

Challenge +10
1. 11	2. 7	3. 18	4. 5	5. 13
6. 4	7. 19	8. 2	9. 15	10. 1
11. 9	12. 14	13. 8	14. 20	15. 16
16. 12	17. 6	18. 17	19. 3	20. 10

Warm-Up +11
1. 17 2. 31 3. 15 4. 28 5. 25
6. 30 7. 19 8. 12 9. 23 10. 21
11. 13 12. 26 13. 20 14. 16 15. 24
16. 22 17. 14 18. 29 19. 18 20. 27

Warm-Up +14
1. 26 2. 20 3. 33 4. 16 5. 29
6. 18 7. 34 8. 21 9. 32 10. 24
11. 31 12. 22 13. 25 14. 15 15. 27
16. 23 17. 17 18. 28 19. 19 20. 30

Warm-Up +17
1. 22 2. 32 3. 27 4. 25 5. 30
6. 33 7. 21 8. 36 9. 19 10. 37
11. 26 12. 18 13. 35 14. 23 15. 28
16. 20 17. 29 18. 24 19. 31 20. 34

Practice +11
1. 16 2. 21 3. 19 4. 30 5. 24
6. 27 7. 14 8. 25 9. 17 10. 22
11. 15 12. 20 13. 23 14. 12 15. 31
16. 29 17. 13 18. 26 19. 18 20. 28

Practice +14
1. 23 2. 25 3. 22 4. 17 5. 32
6. 15 7. 30 8. 19 9. 26 10. 33
11. 21 12. 27 13. 16 14. 20 15. 24
16. 18 17. 28 18. 29 19. 34 20. 31

Practice +17
1. 19 2. 30 3. 37 4. 21 5. 34
6. 23 7. 26 8. 18 9. 36 10. 31
11. 29 12. 24 13. 20 14. 32 15. 27
16. 33 17. 22 18. 28 19. 25 20. 35

Challenge +11
1. 11 2. 19 3. 1 4. 14 5. 8
6. 5 7. 15 8. 7 9. 20 10. 4
11. 18 12. 3 13. 10 14. 16 15. 12
16. 9 17. 6 18. 13 19. 2 20. 17

Challenge +14
1. 16 2. 3 3. 14 4. 6 5. 11
6. 5 7. 9 8. 17 9. 1 10. 19
11. 13 12. 4 13. 10 14. 18 15. 7
16. 12 17. 20 18. 2 19. 8 20. 15

Challenge +17
1. 3 2. 9 3. 17 4. 6 5. 14
6. 11 7. 5 8. 18 9. 13 10. 1
11. 15 12. 2 13. 20 14. 8 15. 19
16. 7 17. 10 18. 16 19. 4 20. 12

Warm-Up +12
1. 15 2. 23 3. 29 4. 18 5. 26
6. 31 7. 19 8. 14 9. 27 10. 32
11. 17 12. 24 13. 21 14. 30 15. 22
16. 13 17. 20 18. 28 19. 25 20. 16

Warm-Up +15
1. 20 2. 31 3. 18 4. 23 5. 25
6. 28 7. 21 8. 33 9. 35 10. 29
11. 26 12. 17 13. 30 14. 24 15. 32
16. 16 17. 34 18. 22 19. 19 20. 27

Warm-Up +18
1. 27 2. 23 3. 30 4. 19 5. 34
6. 32 7. 20 8. 38 9. 25 10. 36
11. 22 12. 35 13. 26 14. 31 15. 28
16. 37 17. 24 18. 29 19. 21 20. 33

Practice +12
1. 30 2. 20 3. 25 4. 16 5. 32
6. 17 7. 24 8. 22 9. 28 10. 13
11. 26 12. 14 13. 29 14. 18 15. 31
16. 19 17. 23 18. 15 19. 27 20. 21

Practice +15
1. 24 2. 18 3. 29 4. 16 5. 23
6. 27 7. 22 8. 31 9. 20 10. 34
11. 21 12. 25 13. 17 14. 28 15. 32
16. 30 17. 19 18. 33 19. 26 20. 35

Practice +18
1. 24 2. 33 3. 20 4. 31 5. 37
6. 25 7. 32 8. 35 9. 21 10. 30
11. 38 12. 29 13. 34 14. 23 15. 36
16. 27 17. 19 18. 28 19. 22 20. 26

Challenge +12
1. 12 2. 19 3. 1 4. 15 5. 5
6. 2 7. 11 8. 6 9. 17 10. 8
11. 10 12. 16 13. 20 14. 13 15. 3
16. 14 17. 9 18. 4 19. 18 20. 7

Challenge +15
1. 11 2. 7 3. 16 4. 3 5. 14
6. 2 7. 17 8. 5 9. 20 10. 9
11. 6 12. 15 13. 10 14. 1 15. 12
16. 13 17. 4 18. 19 19. 8 20. 18

Challenge +18
1. 14 2. 3 3. 18 4. 9 5. 1
6. 5 7. 12 8. 7 9. 20 10. 15
11. 17 12. 2 13. 19 14. 6 15. 11
16. 10 17. 16 18. 4 19. 13 20. 8

Warm-Up +13
1. 19 2. 32 3. 18 4. 24 5. 29
6. 26 7. 17 8. 33 9. 22 10. 15
11. 21 12. 28 13. 14 14. 27 15. 20
16. 16 17. 23 18. 30 19. 25 20. 31

Warm-Up +16
1. 17 2. 30 3. 23 4. 34 5. 21
6. 28 7. 25 8. 20 9. 31 10. 36
11. 22 12. 33 13. 24 14. 26 15. 18
16. 19 17. 32 18. 27 19. 29 20. 35

Warm-Up +19
1. 24 2. 33 3. 26 4. 29 5. 21
6. 20 7. 39 8. 32 9. 28 10. 35
11. 31 12. 22 13. 37 14. 25 15. 30
16. 27 17. 34 18. 38 19. 23 20. 36

Practice +13
1. 24 2. 16 3. 29 4. 33 5. 19
6. 18 7. 21 8. 30 9. 14 10. 25
11. 23 12. 17 13. 32 14. 20 15. 22
16. 27 17. 15 18. 31 19. 28 20. 26

Practice +16
1. 25 2. 19 3. 32 4. 34 5. 21
6. 24 7. 22 8. 28 9. 20 10. 26
11. 30 12. 17 13. 33 14. 23 15. 29
16. 27 17. 36 18. 31 19. 18 20. 35

Practice +19
1. 22 2. 30 3. 27 4. 33 5. 37
6. 25 7. 34 8. 23 9. 38 10. 29
11. 20 12. 31 13. 39 14. 24 15. 35
16. 32 17. 28 18. 36 19. 21 20. 26

Challenge +13
1. 1 2. 9 3. 17 4. 5 5. 13
6. 15 7. 4 8. 19 9. 8 10. 11
11. 12 12. 7 13. 2 14. 14 15. 18
16. 3 17. 16 18. 10 19. 20 20. 6

Challenge +16
1. 14 2. 16 3. 3 4. 19 5. 10
6. 4 7. 11 8. 20 9. 7 10. 15
11. 8 12. 2 13. 18 14. 5 15. 12
16. 13 17. 6 18. 17 19. 9 20. 1

Challenge +19
1. 12 2. 1 3. 16 4. 5 5. 19
6. 3 7. 11 8. 9 9. 14 10. 7
11. 8 12. 15 13. 2 14. 17 15. 13
16. 18 17. 6 18. 20 19. 10 20. 4

Addition Review +1–10
1. 10　2. 11　3. 6　4. 14　5. 12
6. 12　7. 20　8. 7　9. 8　10. 8
11. 9　12. 12　13. 14　14. 6　15. 18
16. 11　17. 13　18. 17　19. 15　20. 10

Challenge –3
1. 4　2. 16　3. 9　4. 13　5. 20
6. 11　7. 18　8. 23　9. 8　10. 21
11. 14　12. 5　13. 22　14. 10　15. 17
16. 19　17. 7　18. 12　19. 15　20. 6

Challenge –6
1. 21　2. 10　3. 23　4. 15　5. 25
6. 14　7. 17　8. 11　9. 8　10. 20
11. 19　12. 7　13. 26　14. 12　15. 18
16. 13　17. 22　18. 9　19. 16　20. 24

Addition Review +11–20
1. 31　2. 30　3. 25　4. 33　5. 28
6. 31　7. 32　8. 36　9. 31　10. 34
11. 24　12. 31　13. 28　14. 35　15. 31
16. 37　17. 29　18. 29　19. 29　20. 35

Warm-Up –4
1. 9　2. 11　3. 3　4. 15　5. 19
6. 2　7. 8　8. 18　9. 4　10. 7
11. 5　12. 16　13. 20　14. 13　15. 17
16. 14　17. 12　18. 6　19. 10　20. 1

Warm-Up –7
1. 9　2. 15　3. 3　4. 18　5. 7
6. 1　7. 8　8. 12　9. 4　10. 17
11. 5　12. 16　13. 20　14. 11　15. 13
16. 14　17. 10　18. 19　19. 6　20. 2

Addition Review +1–20A
1. 15　2. 13　3. 26　4. 10　5. 18
6. 23　7. 19　8. 30　9. 20　10. 14
11. 16　12. 14　13. 16　14. 18　15. 16
16. 17　17. 22　18. 25　19. 17　20. 25

Practice –4
1. 3　2. 6　3. 15　4. 9　5. 18
6. 11　7. 16　8. 1　9. 7　10. 14
11. 2　12. 10　13. 13　14. 5　15. 19
16. 8　17. 4　18. 17　19. 20　20. 12

Practice –7
1. 20　2. 13　3. 18　4. 5　5. 16
6. 11　7. 4　8. 15　9. 8　10. 12
11. 17　12. 6　13. 19　14. 1　15. 10
16. 9　17. 2　18. 7　19. 14　20. 3

Addition Review +1–20B
1. 11　2. 15　3. 22　4. 17　5. 8
6. 23　7. 22　8. 14　9. 27　10. 23
11. 21　12. 18　13. 38　14. 13　15. 18
16. 39　17. 20　18. 28　19. 27　20. 12

Challenge –4
1. 17　2. 11　3. 5　4. 19　5. 8
6. 6　7. 13　8. 22　9. 10　10. 23
11. 20　12. 9　13. 16　14. 21　15. 14
16. 15　17. 24　18. 12　19. 7　20. 18

Challenge –7
1. 27　2. 10　3. 18　4. 22　5. 15
6. 20　7. 8　8. 23　9. 12　10. 25
11. 13　12. 26　13. 16　14. 9　15. 21
16. 24　17. 11　18. 17　19. 14　20. 19

Warm-Up –2
1. 5　2. 13　3. 4　4. 8　5. 15
6. 16　7. 20　8. 12　9. 3　10. 11
11. 7　12. 18　13. 1　14. 9　15. 14
16. 2　17. 10　18. 17　19. 6　20. 19

Warm-Up –5
1. 14　2. 7　3. 18　4. 3　5. 11
6. 5　7. 2　8. 19　9. 9　10. 15
11. 12　12. 8　13. 16　14. 20　15. 6
16. 1　17. 17　18. 10　19. 13　20. 4

Warm-Up –8
1. 10　2. 3　3. 18　4. 7　5. 14
6. 5　7. 8　8. 16　9. 2　10. 11
11. 12　12. 1　13. 19　14. 13　15. 6
16. 15　17. 4　18. 17　19. 20　20. 9

Practice –2
1. 1　2. 7　3. 18　4. 17　5. 12
6. 9　7. 2　8. 13　9. 10　10. 15
11. 19　12. 5　13. 3　14. 16　15. 8
16. 6　17. 11　18. 4　19. 20　20. 14

Practice –5
1. 10　2. 5　3. 18　4. 7　5. 3
6. 14　7. 8　8. 1　9. 15　10. 12
11. 2　12. 11　13. 17　14. 6　15. 19
16. 13　17. 16　18. 4　19. 9　20. 20

Practice –8
1. 20　2. 13　3. 17　4. 11　5. 4
6. 9　7. 2　8. 19　9. 7　10. 16
11. 6　12. 15　13. 10　14. 3　15. 14
16. 18　17. 5　18. 12　19. 1　20. 8

Challenge –2
1. 9　2. 14　3. 5　4. 17　5. 12
6. 16　7. 3　8. 21　9. 10　10. 22
11. 7　12. 19　13. 13　14. 6　15. 8
16. 11　17. 18　18. 4　19. 20　20. 15

Challenge –5
1. 18　2. 13　3. 8　4. 17　5. 22
6. 15　7. 11　8. 19　9. 12　10. 25
11. 10　12. 7　13. 14　14. 24　15. 16
16. 21　17. 20　18. 6　19. 9　20. 23

Challenge –8
1. 24　2. 20　3. 12　4. 28　5. 16
6. 10　7. 18　8. 25　9. 14　10. 22
11. 15　12. 21　13. 9　14. 23　15. 26
16. 27　17. 13　18. 19　19. 17　20. 11

Warm-Up –3
1. 2　2. 11　3. 17　4. 4　5. 8
6. 14　7. 7　8. 20　9. 18　10. 15
11. 9　12. 3　13. 16　14. 12　15. 5
16. 6　17. 13　18. 19　19. 1　20. 10

Warm-Up –6
1. 3　2. 10　3. 19　4. 7　5. 15
6. 12　7. 5　8. 14　9. 11　10. 20
11. 8　12. 1　13. 16　14. 4　15. 18
16. 17　17. 6　18. 13　19. 2　20. 9

Warm-Up –9
1. 2　2. 14　3. 6　4. 17　5. 10
6. 19　7. 8　8. 18　9. 3　10. 16
11. 11　12. 5　13. 20　14. 15　15. 7
16. 13　17. 4　18. 12　19. 9　20. 1

Practice –3
1. 20　2. 9　3. 15　4. 4　5. 12
6. 19　7. 13　8. 1　9. 16　10. 7
11. 3　12. 17　13. 5　14. 10　15. 6
16. 11　17. 8　18. 14　19. 18　20. 2

Practice –6
1. 11　2. 6　3. 14　4. 4　5. 16
6. 8　7. 3　8. 19　9. 9　10. 1
11. 15　12. 20　13. 18　14. 7　15. 12
16. 13　17. 5　18. 10　19. 2　20. 17

Practice –9
1. 6　2. 17　3. 10　4. 18　5. 13
6. 14　7. 7　8. 2　9. 16　10. 19
11. 3　12. 11　13. 8　14. 20　15. 5
16. 15　17. 4　18. 12　19. 1　20. 9

Challenge –9
1. 25 2. 16 3. 22 4. 14 5. 19
6. 12 7. 20 8. 27 9. 11 10. 28
11. 15 12. 24 13. 10 14. 21 15. 17
16. 23 17. 18 18. 29 19. 13 20. 26

Challenge –12
1. 21 2. 26 3. 15 4. 23 5. 31
6. 17 7. 28 8. 19 9. 25 10. 13
11. 30 12. 14 13. 22 14. 18 15. 29
16. 24 17. 20 18. 32 19. 16 20. 27

Challenge –15
1. 16 2. 23 3. 27 4. 21 5. 33
6. 20 7. 25 8. 31 9. 18 10. 29
11. 28 12. 17 13. 35 14. 24 15. 32
16. 34 17. 22 18. 30 19. 19 20. 26

Warm-Up –10
1. 15 2. 9 3. 20 4. 5 5. 12
6. 17 7. 13 8. 4 9. 10 10. 2
11. 19 12. 8 13. 1 14. 14 15. 16
16. 6 17. 11 18. 7 19. 18 20. 3

Warm-Up –13
1. 8 2. 1 3. 17 4. 12 5. 5
6. 4 7. 16 8. 9 9. 19 10. 14
11. 11 12. 20 13. 2 14. 18 15. 7
16. 13 17. 6 18. 10 19. 15 20. 3

Warm-Up –16
1. 4 2. 12 3. 20 4. 6 5. 18
6. 8 7. 1 8. 16 9. 14 10. 10
11. 17 12. 5 13. 13 14. 2 15. 19
16. 11 17. 3 18. 7 19. 15 20. 9

Practice –10
1. 11 2. 4 3. 10 4. 7 5. 12
6. 3 7. 15 8. 19 9. 5 10. 16
11. 8 12. 17 13. 2 14. 14 15. 9
16. 13 17. 6 18. 20 19. 18 20. 1

Practice –13
1. 15 2. 17 3. 12 4. 18 5. 20
6. 3 7. 13 8. 6 9. 16 10. 19
11. 14 12. 11 13. 2 14. 7 15. 8
16. 1 17. 5 18. 10 19. 4 20. 9

Practice –16
1. 5 2. 12 3. 16 4. 10 5. 2
6. 1 7. 14 8. 7 9. 18 10. 20
11. 9 12. 19 13. 3 14. 17 15. 8
16. 6 17. 11 18. 15 19. 4 20. 13

Challenge –10
1. 22 2. 19 3. 27 4. 15 5. 29
6. 14 7. 26 8. 17 9. 12 10. 21
11. 28 12. 11 13. 20 14. 18 15. 24
16. 25 17. 16 18. 23 19. 30 20. 13

Challenge –13
1. 18 2. 30 3. 22 4. 16 5. 24
6. 27 7. 32 8. 14 9. 31 10. 20
11. 21 12. 29 13. 17 14. 26 15. 25
16. 15 17. 33 18. 23 19. 19 20. 28

Challenge –16
1. 20 2. 29 3. 23 4. 31 5. 17
6. 25 7. 33 8. 21 9. 27 10. 36
11. 30 12. 18 13. 24 14. 32 15. 35
16. 34 17. 22 18. 26 19. 19 20. 28

Warm-Up –11
1. 12 2. 7 3. 13 4. 19 5. 5
6. 16 7. 4 8. 1 9. 10 10. 17
11. 6 12. 15 13. 9 14. 18 15. 3
16. 2 17. 11 18. 20 19. 8 20. 14

Warm-Up –14
1. 3 2. 9 3. 15 4. 6 5. 12
6. 11 7. 5 8. 19 9. 1 10. 17
11. 14 12. 20 13. 8 14. 16 15. 4
16. 7 17. 18 18. 2 19. 13 20. 10

Warm-Up –17
1. 12 2. 1 3. 18 4. 6 5. 10
6. 8 7. 16 8. 4 9. 20 10. 14
11. 5 12. 17 13. 13 14. 2 15. 11
16. 3 17. 9 18. 15 19. 19 20. 7

Practice –11
1. 19 2. 12 3. 5 4. 14 5. 8
6. 1 7. 7 8. 20 9. 4 10. 15
11. 16 12. 10 13. 9 14. 13 15. 2
16. 18 17. 17 18. 6 19. 3 20. 11

Practice –14
1. 16 2. 13 3. 15 4. 6 5. 12
6. 9 7. 4 8. 2 9. 1 10. 3
11. 5 12. 14 13. 20 14. 8 15. 18
16. 11 17. 7 18. 10 19. 19 20. 17

Practice –17
1. 5 2. 13 3. 10 4. 17 5. 3
6. 2 7. 16 8. 7 9. 20 10. 14
11. 19 12. 6 13. 12 14. 1 15. 9
16. 8 17. 15 18. 4 19. 11 20. 18

Challenge –11
1. 12 2. 26 3. 21 4. 15 5. 24
6. 28 7. 14 8. 23 9. 18 10. 30
11. 16 12. 27 13. 31 14. 19 15. 13
16. 25 17. 22 18. 17 19. 29 20. 20

Challenge –14
1. 16 2. 23 3. 29 4. 33 5. 20
6. 27 7. 19 8. 25 9. 32 10. 17
11. 21 12. 28 13. 15 14. 24 15. 34
16. 18 17. 31 18. 26 19. 22 20. 30

Challenge –17
1. 24 2. 30 3. 22 4. 33 5. 27
6. 21 7. 25 8. 34 9. 20 10. 31
11. 37 12. 28 13. 19 14. 32 15. 36
16. 18 17. 26 18. 35 19. 23 20. 29

Warm-Up –12
1. 4 2. 18 3. 16 4. 11 5. 3
6. 10 7. 2 8. 20 9. 5 10. 13
11. 7 12. 14 13. 9 14. 1 15. 17
16. 12 17. 19 18. 6 19. 8 20. 15

Warm-Up –15
1. 11 2. 6 3. 14 4. 3 5. 9
6. 2 7. 17 8. 5 9. 20 10. 13
11. 18 12. 4 13. 12 14. 8 15. 15
16. 7 17. 16 18. 10 19. 19 20. 1

Warm-Up –18
1. 4 2. 12 3. 8 4. 16 5. 9
6. 15 7. 7 8. 20 9. 3 10. 11
11. 19 12. 1 13. 18 14. 13 15. 6
16. 5 17. 10 18. 17 19. 2 20. 14

Practice –12
1. 20 2. 6 3. 12 4. 4 5. 15
6. 3 7. 9 8. 16 9. 5 10. 10
11. 18 12. 2 13. 17 14. 11 15. 8
16. 7 17. 13 18. 14 19. 1 20. 19

Practice –15
1. 8 2. 5 3. 15 4. 18 5. 7
6. 16 7. 2 8. 17 9. 4 10. 13
11. 11 12. 9 13. 14 14. 19 15. 12
16. 6 17. 3 18. 20 19. 10 20. 1

Practice –18
1. 1 2. 19 3. 4 4. 10 5. 13
6. 12 7. 8 8. 18 9. 6 10. 16
11. 14 12. 3 13. 20 14. 11 15. 7
16. 5 17. 17 18. 9 19. 15 20. 2

Challenge –18

1. 29	2. 22	3. 37	4. 25	5. 31
6. 23	7. 26	8. 32	9. 35	10. 20
11. 34	12. 30	13. 21	14. 38	15. 27
16. 36	17. 19	18. 28	19. 33	20. 24

Warm-Up –19

1. 4	2. 11	3. 18	4. 5	5. 16
6. 8	7. 20	8. 1	9. 14	10. 7
11. 2	12. 15	13. 9	14. 12	15. 19
16. 13	17. 6	18. 17	19. 3	20. 10

Practice –19

1. 13	2. 5	3. 16	4. 1	5. 20
6. 9	7. 18	8. 3	9. 14	10. 6
11. 2	12. 11	13. 17	14. 10	15. 15
16. 8	17. 12	18. 19	19. 4	20. 7

Challenge –19

1. 28	2. 36	3. 31	4. 25	5. 39
6. 30	7. 23	8. 35	9. 21	10. 33
11. 27	12. 34	13. 20	14. 29	15. 26
16. 22	17. 38	18. 32	19. 24	20. 37

Subtraction Review 1–20A

1. 3	2. 7	3. 5	4. 5	5. 4
6. 2	7. 14	8. 1	9. 4	10. 13
11. 12	12. 6	13. 7	14. 8	15. 11
16. 5	17. 12	18. 1	19. 12	20. 12

Subtraction Review 1–20B

1. 3	2. 5	3. 1	4. 3	5. 4
6. 7	7. 7	8. 5	9. 2	10. 8
11. 5	12. 5	13. 8	14. 13	15. 6
16. 13	17. 12	18. 10	19. 5	20. 9

Addition and Subtraction Review
1–20A

1. 14	2. 18	3. 14	4. 3	5. 5
6. 6	7. 16	8. 20	9. 10	10. 37
11. 2	12. 14	13. 11	14. 32	15. 6
16. 8	17. 18	18. 22	19. 6	20. 16

Addition and Subtraction Review
1–20B

1. 10	2. 12	3. 16	4. 24	5. 7
6. 13	7. 28	8. 3	9. 12	10. 9
11. 8	12. 38	13. 13	14. 25	15. 18
16. 13	17. 22	18. 6	19. 24	20. 22

Addition and Subtraction Review
1–20C

1. 10	2. 21	3. 6	4. 12	5. 25
6. 21	7. 14	8. 20	9. 12	10. 33
11. 4	12. 36	13. 17	14. 12	15. 6
16. 34	17. 4	18. 23	19. 21	20. 14

Addition and Subtraction Review
1–20D

1. 24	2. 4	3. 32	4. 13	5. 23
6. 13	7. 36	8. 6	9. 8	10. 12
11. 15	12. 4	13. 20	14. 16	15. 19
16. 17	17. 24	18. 3	19. 33	20. 14